Gilera
The Complete Story

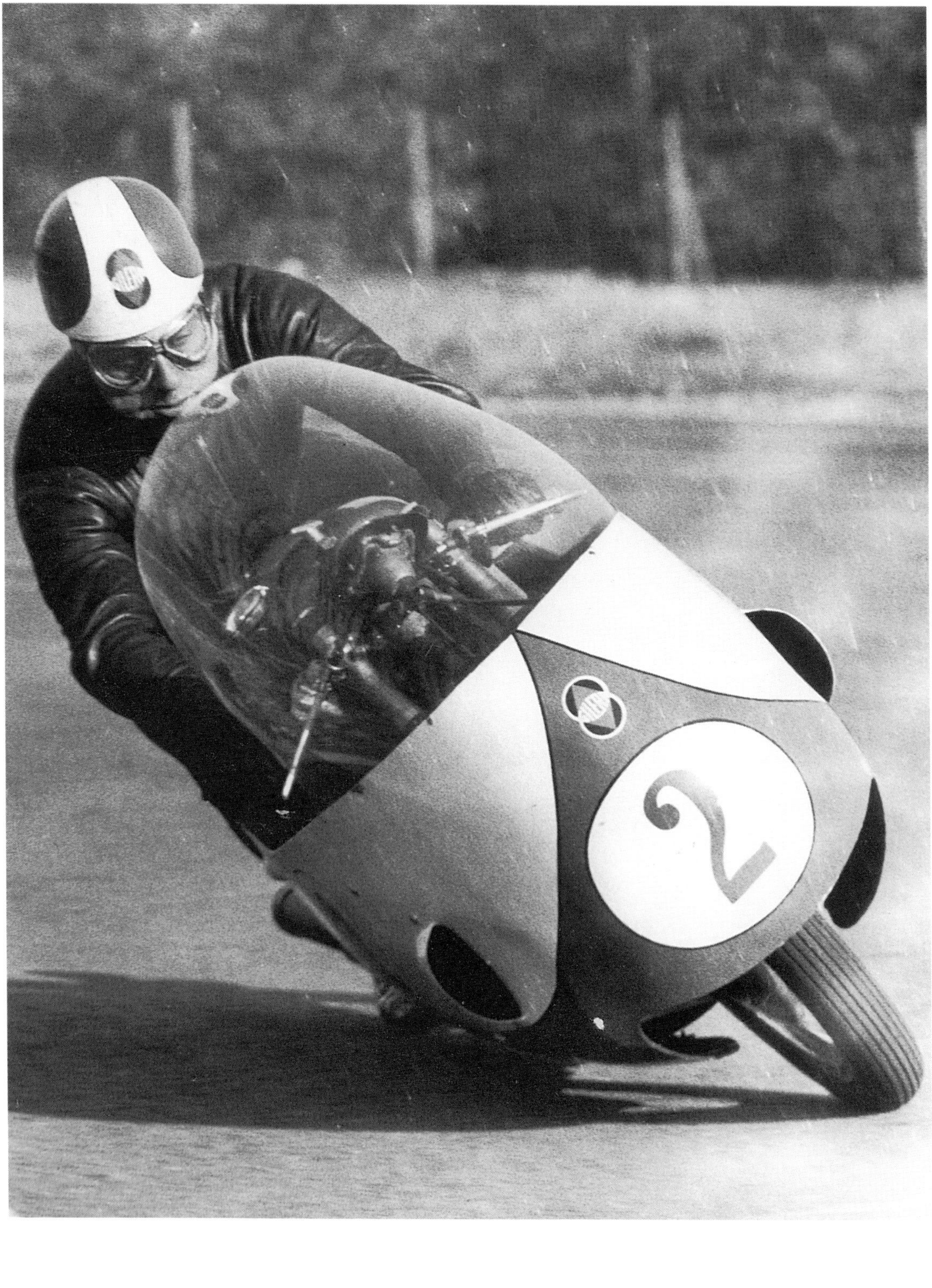

GILERA
The Complete Story

Mick Walker

The Crowood Press

First published in 2000 by
The Crowood Press Ltd
Ramsbury, Marlborough
Wiltshire SN8 2HR

© Mick Walker 2000

All rights reserved. No part of this publication may be reproduced or transmitted in any form or by any means, electronic or mechanical, including photocopy, recording, or any information storage and retrieval system, without permission in writing from the publishers.

British Library Cataloguing-in-Publication Data
A catalogue record for this book is available from the British Library.

ISBN 1 86126 333 3

Dedication
To the memory of Gary Walker, a much missed son and friend.

Typeface used: New Century Schoolbook.

Typeset by Jean Cussons Typesetting
Diss, Norfolk

Printed and bound by The Bath Press

Contents

	Acknowledgements	6
1	In the Beginning	7
2	Saturno	24
3	Grand Prix Glory	41
4	Record Breaking	83
5	Decline and Fall	92
6	Piaggio	112
7	Dirt Bikes	123
8	Two-Strokes – the New Breed	138
9	Four-Strokes – the New Breed	152
10	End of the Road	165
11	Technical Appraisal	174
12	Rebirth	180
	Index	190

Acknowledgements

Although MV Agusta ultimately won more races and world titles, it was Gilera who were really responsible for the development of the modern across-the-frame four-cylinder engine. Even the mighty Japanese have used the same formula to create the majority of their large capacity machines over the last thirty years, beginning with the Honda CB750 of 1969. The importance, therefore, of Gilera to the overall history of the motorcycle cannot be overestimated.

My own first personal experience of the four-cylinder Gilera racing machine came at Silverstone in April 1963, when I was there to witness at first hand the vanguard outing on British soil of the Scuderia Duke team of Derek Minter and John Hartle. I can still remember that day clearly – the high wave of excitement everyone felt in the stands opposite the pits as the two, four-cylinder Gileras were brought to the start line for the 500cc event – even if Phil Read on a privately entered Manx Norton single hounded Minter and Hartle throughout the race, even though the Gilera's speed eventually paid off.

However, I've had a lot of other Gilera experiences over the years, including having ridden many of the marque's best. These include both the original and *nuovo* Saturnos, the chance to ride a Piuma racer around Mallory Park Circuit, a visit to the Arcore factory in 1988 and, finally, meeting the many famous Gilera stars of the past, including Duke Masetti and Caldarella at the Centennial TT, Assan, in May 1998.

As with my earlier books, I have received a huge amount of help and support in the preparation of *Gilera – the Complete Story*. Those involved include Guiseppe Tranchina, Constantino Sambug and David Champion of Piaggio UK, Shirley Pattison of Promark, racer Sally Kelly for the use of her immaculate Piuma, Doug Jackson, Gerald Gilligen and last, but by no means least, Dave and Mark Kay.

Besides the factory and my own collection, illustrations came from Doug Jackson, Ian Welsh, Vic Bates and David Campion, plus Gilera and Piaggio themselves.

In closing I have to say that I really enjoyed penning the Gilera story and, having previously written about MV Agusta, I feel this is almost the follow-up volume to that book as the two marques were so closely linked in many crucial areas down the years.

Mick Walker
February 2000
Wisbech, Cambridgeshire

1 In the Beginning

Any history of Gilera is duty bound to pay great attention to an engine which changed the course of motorcycling history itself: the across-the-frame, four-cylinder. Today this layout is largely associated with the Japanese, but, as this book will relate, it was in another country and in another time that the concept was born.

We must go back to Italy and the year 1923 to find its origins. It was then that two young engineers from Rome, Carlo Gianini and Piero Remor, designed and built a prototype four-cylinder, single overhead camshaft engine with a capacity of 490cc (51 × 60mm), which was driven by a train of gears set between two pairs of cylinders.

THE GRB FOUR

The following year, joined by another motorcycle enthusiast in Rome, Count Giovanni Bonmartini, they produced their first complete bike – the GRB, named after the initials of the trio. Development of this machine continued until 1928, by which time it was producing around 28bhp at 6,000rpm.

An air-cooled unit until then, the engine was modified by having the exhaust zone and the area around each sparking plug cooled by water. Power output was boosted to 32bhp at 6,500rpm.

The original 1923 GRB design, with single ohc actuated by a shaft in front of the four cylinders set across the frame. The 490.276 (51 × 60mm) unit developed 28bhp at 6,000rpm, but was not raced.

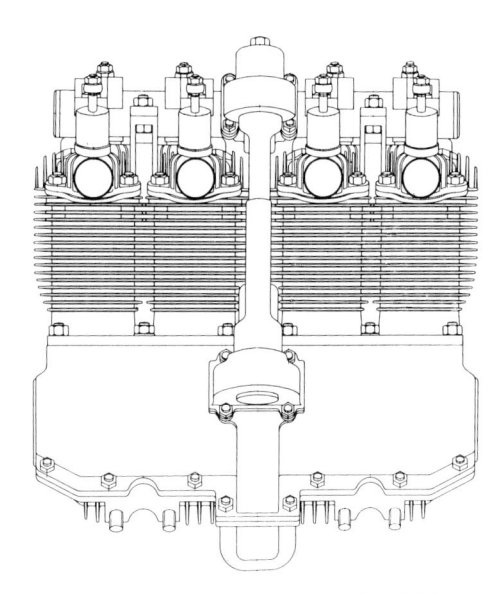

Frontal view drawing of the air-cooled GRB four-cylinder engine.

In the Beginning

The most significant step from the GRB to the OPRA of 1927 was in the drive to the ohc. A train of spur gears was introduced, a feature retained in future development of the power unit, even when it was taken over by Gilera.

Two years further down the track and the whole GRB project was in grave danger of folding through a lack of financial resources. Desperate attempts were made by the partners to persuade leading factories, abroad as well as in Italy, to take on board their handiwork, but all were to prove fruitless. So the CNA (Compagnia Nazionale Aeronautica) of Rome, owned by Count Bonmartini, acquired the GRB assets and design. When this took place, Ing. Gianini switched to CNA, but for a considerable time he was involved purely in working on engines for light aircraft use.

THE RONDINE

It was not until 1934 that Count Bonmartini's enthusiasm for motorcycles found fresh expression in a redesigned version of the original four-cylinder machine. This was known as the Rondine (fast-flying Swallow). Ing. Remor once again became associated with Gianini in its development and working with them was another engineer, Piero Taruffi, who was also a rider of considerable note.

The motorcycle that they conceived featured twin overhead camshafts (still with a central gear drive), inclined cylinders which were totally water-cooled, and a supercharger. Power output was a sensational 86bhp at 9,000rpm. Equipped with a four-speed gearbox featuring positive-stop foot-change, it was to achieve a brilliant début at Tripoli Autodrome, where Taruffi and Amilcare Rossetti scored an emphatic one–two, beating the Guzzi and Norton factory teams.

In the Beginning

Count Giovanni Bonmartini (in hat) talking to Pierre Taruffi who is astride the Rondine (34) prior to the start of a race at Rome in 1934. With water-cooling, supercharging twin ohc and the change from vertical to inclined (45 degrees) cylinders, there had been dramatic technical changes from the OPRA model. New bore and stroke measurements of 52 × 58mm, gave 492.692cc.

The following year, with a partially streamlined Rondine, Taruffi set a number of world speed records (*see* Chapter 4). In that same year, however, Count Bonmartini sold his business to the vast Caproni aeronautical enterprise. As Caproni had no interest in bikes at that time, it looked around for someone who would be interested in acquiring the Rondine portion of the CNA operation.

Also involved in the early development of Gilera's four-cylinder racer was Ing. Carlo Gianini (centre). Flanking him in this 1935 shot are Rondine riders Amilcare Rossetti (left) and Piero Taruffi. The Rondine also pioneered dolphin-style streamlining and the lateral frame concept.

9

In the Beginning

GILERA ARRIVES ON THE SCENE

At this point Giuseppe Gilera arrived on the scene. He was born into a working-class family in a small village near Milan in 1887. From an early age he had been fascinated by any form of mechanical transportation. When only fifteen, the young Gilera entered employment with the Bianchi factory in Milan, where he gained his first practical experience.

Moving first to the Moto-Reve works and then to the well-known engineering concern of Bucher and Zeda, Giuseppe Gilera quickly built up a vast store of knowledge which was to serve him in good stead.

By now, the ambitious young engineer had also taken up road racing and, in particular, gained recognition for his excellent showing in hill climb events. However, the speed events always came second to his burning desire to become a motorcycle manufacturer in his own right. By 1909, at the age of twenty-two, he was ready to take the plunge.

THE 317CC SINGLE

His first model was a 317cc (67 × 90mm) four-stroke single featuring an ohv (overhead valve) engine with both inlet and exhaust valves operated mechanically – quite a rare layout at the time. Next came a V-twin and, thereafter, the first of the famous 500cc ohv singles – forerunners of the legendary post-war Saturno series which made such a mark, not only on the road, but also in trials, motocross and racing.

Right from the start, Giuseppe Gilera appreciated the importance of publicity gained through sporting successes. The marque's first ever victory came at the Cremona circuit in 1912.

Gilera's first motorcycle, a 317cc (67 × 90mm) long-stroke ohv single, circa 1909. Giuseppe Gilera was then 22 years old.

In the Beginning

> **317 Single (1909–1912)**
>
> | | |
> |---|---|
> | Engine: | Air-cooled, single-cylinder, four-stroke with both inlet and exhaust valves operated mechanically |
> | Bore: | 67mm |
> | Stroke: | 90mm |
> | Displacement: | 317.3cc |
> | Compression ratio: | 5:1 |
> | Maximum power (at crank): | 7bhp @ 3,000rpm |
> | Lubrication: | Wet sump, pump |
> | Ignition: | Magneto, UH |
> | Fuel system: | Binks carburettor |
> | Primary drive: | Not applicable – power transmitted direct from crankshaft to rear wheel by belt |
> | Final drive: | Belt, with pedal assistance if required |
> | Gearbox: | Single speed |
> | Frame: | Steel tubular, cycle type |
> | Front suspension: | Cycle fork |
> | Rear suspension: | Rigid |
> | Front brake: | None |
> | Rear brake: | Block |
> | Wheels: | 26in front and rear |
> | Tyres: | 26 × 1.75in front and rear |
> | Wheelbase: | 47in (1,200mm) |
> | Dry weight: | 165lb (75kg) |
> | Fuel tank capacity: | 175imp. gal (8ltr) |
> | Top speed: | 65mph (105km/h) |

THE ARCORE FACTORY

After the Great War, the demand for motorcycles accelerated rapidly and Giuseppe Gilera decided to establish a new factory at Arcore, on the main road from Milan to Lecco. This was only a few kilometres from Monza Park, where the famous autodrome was to be constructed.

His first product was the 3.5hp Turismo built in 1920 with 498.76cc (84 × 90mm) ioe (inlet over exhaust) single-cylinder configuration. Two versions were built; one with 8bhp at 3,500rpm, the other with 10bhp at 3,800rpm and a maximum speed of around 75mph (120km/h). There was also a 350 Super Sport (1925–1928) with a 346.350cc (70 × 90mm) engine. The smaller SS offered its owner 10bhp at 4,000rpm at 69mph (111km/h).

346.35cc (70 × 90mm) inlet-over-exhaust 350SS single; produced between 1925 and 1928.

500 Super Sport (1923–1928)

Engine:	Air-cooled, single-cylinder, four-stroke ioe (inlet-over-exhaust)
Bore:	84mm
Stroke:	90mm
Displacement:	498.76cc
Compression ratio:	5:1
Maximum power (at crank):	12bhp @ 3,800rpm
Lubrication:	Wet sump
Ignition:	Magneto, Bosch or Marelli AT
Fuel system:	Binks carburettor
Primary drive:	Chain
Final drive:	Chain
Gearbox:	Three-speed
Frame:	Tubular steel
Front suspension:	Girder forks, with central spring
Rear suspension:	Rigid
Front brake:	None
Rear brake:	Drum
Wheels:	21in front and rear
Tyres:	3.00 × 21 front and rear
Wheelbase:	53.5in (1,360mm)
Dry weight:	275.5lb (125kg)
Fuel tank capacity:	2.2imp. gal (10ltr)
Top speed:	75mph (120km/h)

OVERHEAD VALVE SINGLE

In 1929 all these models were replaced by a new ohv version of the same basic motor, the Gran Sport, which was manufactured until the end of 1931.

Throughout the 1920s and early 1930s, Gilera not only grew into one of the country's largest factories, but gained considerable success in sporting events such as trials and long-distance races.

LUIGI GILERA

Giuseppe Gilera's brother Luigi was a member of the winning Italian Trophy team in the 1930 ISDT, staged around Grenoble, France. This success was to be repeated in 1931, when the event was held in Italy. In that year all the Italian riders were Gilera mounted. Luigi Gilera competed on a sidecar outfit, the engine being one of the new 500-class ohv singles. Other well known Gilera sidecar trials aces of the era included Miro Maffeis and Rosolino Grana.

FOUR-CYLINDER NEWS

However, Gilera's ultimate goal was the International Grand Prix racing scene and, at the beginning of 1936, the chance to acquire the Rondine project was heaven-sent.

Within a year, the machine had been totally updated as the result of a joint team effort by Remor, Taruffi and Giuseppe Gilera himself. Its rigid, pressed-steel frame had been superseded by a tubular device incorporating swinging-arm rear suspension.

In the Beginning

Brother Luigi was a notable rider, he is seen here (on a Gilera of course!) during the speed test section of the 1930 ISDT.

350 Gran Sport (1929–1931)	
Engine:	Air-cooled, single-cylinder, four-stroke side valve
Bore:	70mm
Stroke:	90mm
Displacement:	346.350cc
Compression ratio:	5:1
Maximum power (at crank):	10bhp @ 4,000rpm
Lubrication:	Wet sump
Ignition:	Magneto, Bosch or Marelli AT
Fuel system:	Amac carburettor
Primary drive:	Chain
Final drive:	Chain
Gearbox:	Three-speed
Frame:	Full cradle, steel
Front suspension:	Girder forks, with central spring
Rear suspension:	Rigid
Front brake:	Drum, steel
Rear brake:	Drum, steel
Wheels:	20in front and rear
Tyres:	3.00 × 26 front and rear
Wheelbase:	53.5in (1,360mm)
Dry weight:	275.5lb (125kg)
Fuel tank capacity:	2.5imp. gal (11ltr)
Top speed:	75mph (120km/h)

In the Beginning

	500 Sei Giorni (1931–1934)
Engine:	Air-cooled, single-cylinder, four-stroke, side valve
Bore:	84mm
Stroke:	90mm
Displacement:	498.76cc
Compression ratio:	5:1
Maximum power (at crank):	13.5bhp @ 5,000rpm
Lubrication:	Wet sump
Ignition:	Magneto, Bosch
Fuel system:	Binks carburettor
Primary drive:	Duplex chain
Final drive:	Chain
Gearbox:	Four-speed
Frame:	Tubular steel
Front suspension:	Girder forks, central spring
Rear suspension:	Rigid
Front brake:	Drum, steel
Rear brake:	Drum, steel
Wheels:	9in front and rear
Tyres:	3.25 × 19 front and rear
Wheelbase:	15in (380mm)
Dry weight:	320lb (145kg)
Fuel tank capacity:	2.2imp. gal (10ltr)
Top speed:	72mph (115km/h)

The four-cylinder was raced under the Gilera banner from 1936. This photograph, taken in 1939, shows the machine with its bodywork removed.

In the Beginning

Dating from 1937, this photograph shows the rig-frame sidecar outfit ridden by Luigi Gilera in both long-distance trials and road-racing events.

175 Sirio (1931–1936)	
Engine:	Air-cooled, single-cylinder, four-stroke, side valve
Bore:	64mm
Stroke:	68mm
Displacement:	218.7cc
Compression ratio:	4.5:1
Maximum power (at crank):	1931 5bhp @ 4,200rpm; 1936 6bhp @ 4,000rpm
Lubrication:	Wet sump
Ignition:	Magneto, Bosch
Fuel system:	Binks carburettor
Primary drive:	Duplex chain
Final drive:	Chain
Gearbox:	Four-speed
Frame:	Tubular steel
Front suspension:	Girder forks, central spring
Rear suspension:	Rigid
Front brake:	Drum, steel
Rear brake:	Drum, steel
Wheels:	19in front and rear
Tyres:	3.00 × 19 front and rear
Wheelbase:	52in (1,330mm)
Dry weight:	220lb (100kg)
Fuel tank capacity:	1.8imp. gal (8ltr)
Top speed:	1931 47mph (75km/h); 1936 50mph (80km/h)

Note: A development of this machine with a 247.7cc (64 × 77mm) engine was built between 1937 and 1941, coded L or LE.

500 Tre Valvole (1933)

Engine:	Air-cooled, single-cylinder, four-stroke, ohv, with twin-port, 3-valve cylinder head
Bore:	84mm
Stroke:	90mm
Displacement:	498.76cc
Compression ratio:	5:1
Maximum power (at crank):	20bhp @ 4,500rpm
Lubrication:	Dry sump
Ignition:	Magneto, Bosch
Fuel system:	Dell'Orto MD271 carburettor
Primary drive:	Chain
Final drive:	Chain
Gearbox:	Four-speed
Frame:	Tubular steel
Front suspension:	Girder fork, central spring
Rear suspension:	Rigid
Front brake:	Drum, steel
Rear brake:	Drum, steel
Wheels:	19in front and rear
Tyres:	3.25 × 19 front and rear
Wheelbase:	53.5in (1,360mm)
Dry weight:	298lb (135kg)
Fuel tank capacity:	2.8imp. gal (13ltr)
Top speed:	79 mph (125km/h)

THE 1939 SEASON

For a short period in 1937 Taruffi became the fastest man in the world on two wheels, with a speed of just over 170mph (275km/h) (*see* Chapter 4). In racing, the four was slowly developing until by 1939 it was a serious contender for championship honours. The 492.692cc (52 × 58mm) four-cylinder, water-cooled, dohc engine featured a small high-speed, Roots-type supercharger which forced the fuel mixture into the cylinders via a balance chamber, the purpose of which was to keep the induction-pipe pressure constant and cool the mixture before it was fed to the cylinders.

Running on a compression ratio of no less than 16:1, it was capable of almost 140mph (225km/h). Primary drive was by gear, from a four-speed gearbox. This impressive machine tipped the scales at 397lb (180kg) dry, and the centre of gravity was exceptionally low, thanks to the steeply inclined cylinder banks.

During the final few weeks before the start of the Second World War, (which effectively ended racing for some six years on the international scene), Gilera, in the shape of Dorini Serafini, pushed hard for victory. After falling while leading comfortably in the Dutch TT, as well as finishing second in Belgium, Serafini won the final three races before the war called a halt to the proceedings.

In the Beginning

Four cylinder 500 Supercharger (1937–1946)

Engine:	Water-cooled, four-cylinder, four-stroke, dohc, across-the-frame-four, with steeply inclined cylinders and supercharging
Bore:	52mm
Stroke:	58mm
Displacement:	492.692cc
Compression ratio:	16:1
Maximum power (at crank):	75bhp @ 9,000rpm
Lubrication:	Dry sump
Ignition:	Magneto, Bosch
Fuel system:	Weber ASB compressor with Roots supercharger
Primary drive:	Gear
Final drive:	Chain
Gearbox:	Four-speed
Frame:	Tubular steel
Front suspension:	Blade-type with central spring
Rear suspension:	Boxed horizontal, with friction adjusters
Front brake:	Drum
Rear brake:	Drum
Wheels:	Front 21in; Rear 20in
Tyres:	Front 3.00 × 21; Rear 3.25 × 20
Wheelbase:	59in (1,500mm)
Dry weight:	397lb (180kg)
Fuel tank capacity:	4.8imp. gal (22ltr)
Top speed:	137.5mph (220km/h)

Piero Taruffi (second from right) acted as Gilera's racing manager in 1939, when Dorino Serafina, riding the supercharged four, won the European Championship against the might of Germany and Britain.

In the Beginning

The first of these was the Swedish GP which was watched by 150,000 spectators in early August. Gilera's main rival BMW suffered a major setback when team leader Georg Meier crashed, injuring his back severely enough for him to be ruled out for the remainder of the season. With Meier out of contention, Serafini and newcomer Silvio Vailati made it a Gilera one–two.

A week later Serafini's victory in the German GP was even sweeter. The team then ventured to Ulster for what was not only billed as 'The Fastest European Road Race', but was also the Grand Prix of Europe that year.

Serafini won at an average of 97.85mph (157km/h), over a mile an hour faster than his average for the Swedish event. For the first time the Ulster saw a 100mph (160km/h) lap, but this was not achieved by Gilera; instead the honour went to a British rider (Walter Rusk) and a British bike (the AJS supercharged four); however, the British pairing were unable to last the 246 mile (395km) distance.

Gilera's second string, Silvio Vailati, in action during the 1939 German Grand Prix at the famous Sachsenring circuit.

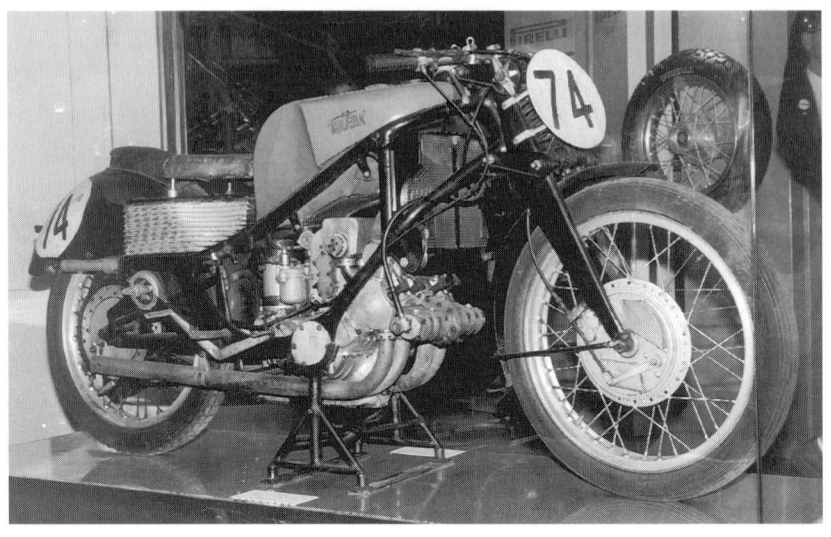

One of the 1939 supercharged Gilera fours on display at the International Milan Show, November 1987.

In the Beginning

With war being declared within days of the Ulster race, the next round of the European Grand Prix (the Italian) never took place. However, as Italy didn't enter the war until June 1940, racing and development continued at Gilera.

One such project was a 250 four. This air-cooled device was destined never to be built, but it did act as a basis for the post-war, air-cooled, four-cylinder range Gileras which were built in both 350 and 500cc engine sizes.

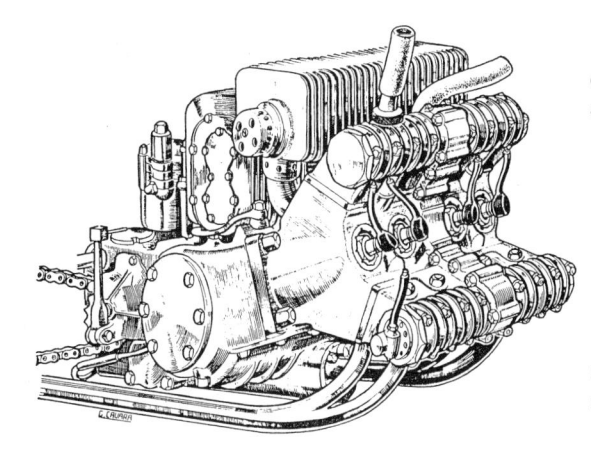

(Left) *The 1939 Gilera four-cylinder engine with Roots blower and water cooling.*

(Below) *Original factory engineering drawing of pre-war four.*

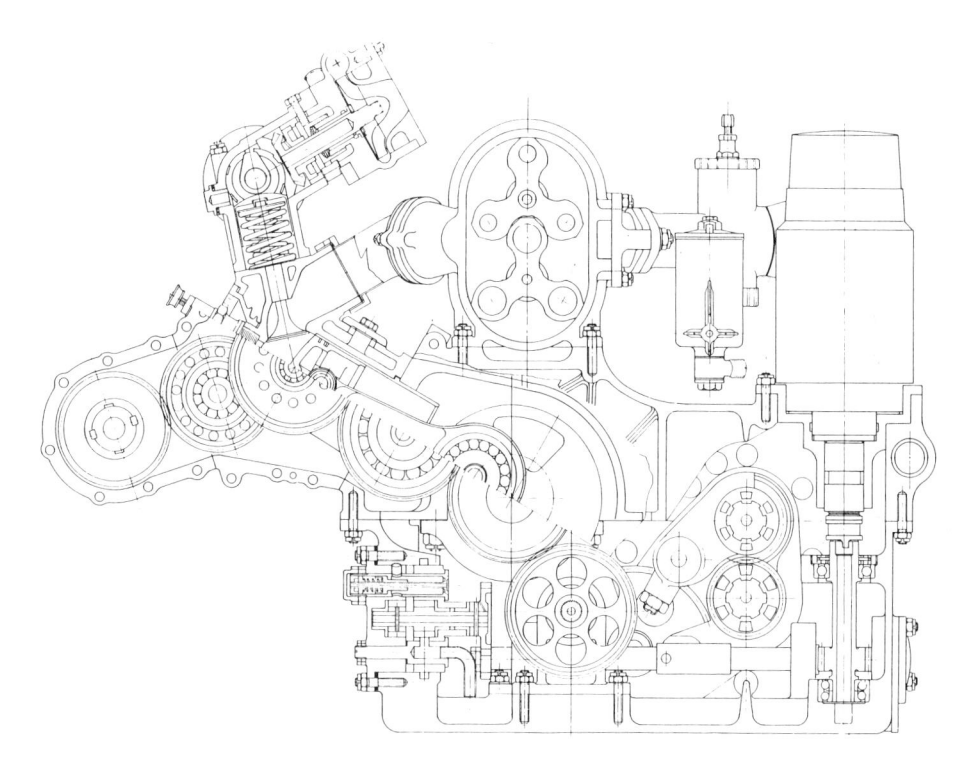

In the Beginning

PRODUCTION ROADSTERS

On the roadster front, besides the models mentioned earlier, Gilera also built the L and SS (350 and 500) from 1931–1934; the *Sei Giorni* (Six Days) 500 1931–1934; the L-LE 500 (1936–1941); the 175 (known as the Sirio) (1931–1936); the L-LE 250 (1937–1941); the Marte – a shaft-drive military machine (1941–1946); the Tre Valvole 500 (1933 only); the Bitubo VT 500 (1934 only) and finally its most popular 'V' line, built from 1935 through to 1941. This range, all ohv 498.76cc singles, consisted of VT, VTGS, VTE and VTGSE. In many ways these were the forerunners to the famous Saturno (*see* Chapter 2) which first appeared as a prototype in 1939, but was not actually manufactured until the post-war period.

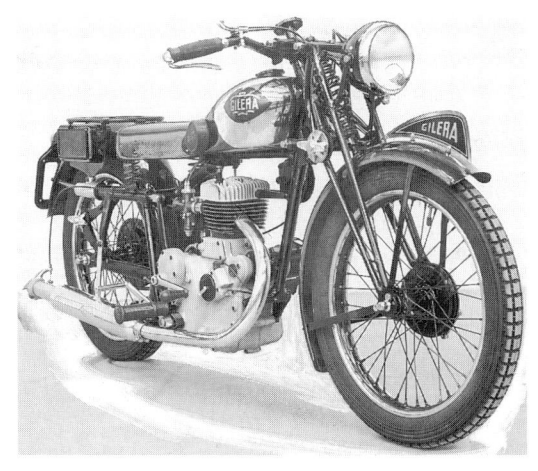

Typical Gilera roadster single of the 1930s. The 247.7cc (64 × 77mm) LE side-valve model. Three-speed foot change gearbox, 9bhp at 4,800rpm. Also sold as the L with rigid frame.

Gilera's home for a large part of its existence was this factory complex at Arcore near the Monza race circuit, Milan.

Another source of revenue for the Arcore factory (like its great rival Moto Guzzi) was the *moto carri* (motorcycle truck). Gilera built a whole series of these, beginning in 1936; 250 (Urano) and 500/600 (Gigante), all with side-valve motors. They were superseded by the Mercurio (built in 500 and 600cc engine sizes) which made its bow in 1940 and ran, little changed, until as late as 1963.

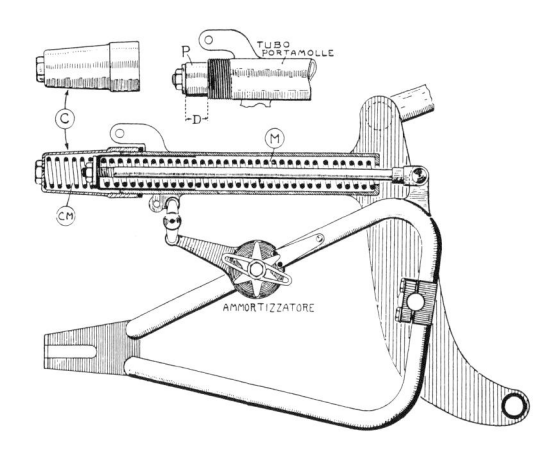

Rear suspension design as used by several Gilera models during the late 1930s.

GILERA ECHO NORTON

In many ways Gilera echoed Norton not only with its racing exploits, but also with its military bikes. Just as the British marque raced the Manx, yet went to war with its plodding 16H side-valves, so did Gilera with the equally gutless LTE model. Running on a modest 4.5:1 compression, this 500 produced a mere 10bhp at 3,400rpm. But the Italian marque did have the much improved Marte, with not only shaft drive, but also a light alloy cylinder head, 5:1 compression ratio and increased power output of 14bhp at 4,800rpm.

There was a sidecar variant with drive to the third wheel without loss of rear suspension; in this Gilera were unique amongst the ranks of military bike builders on both sides during the Second World War. The company also supplied military versions of its *moto carri* load carriers; like their civilian counterparts these were based on a motorcycle back to the saddle, plus a rear axle and load-carrying chassis. On this chassis all types of body were offered to carry a wide range of stores equipment or even to act as a gun platform. Built in either 500 or 600cc side-valve forms or a 500cc ohv unit, they were basically military adaptations of the Gigante and Mercurio respectively.

A consignment of Mercurio Motocarro (three-wheel truck) and Saturno engines at the plant in the immediate post-war period.

In the Beginning

The Mercurio three-wheeler was built from 1940 until as late as 1963. This is one of the first constructed. It was available with either 498.76cc or 582cc ohv engines.

WAR BREAKS OUT

However, for the Italians – or at least the vast majority – the war was simply an interruption of the far more important things in life; namely, women, wine, food and, of course, racing. It is perhaps worth remembering that, as the Nazi hordes swept across Western Europe during May 1940, in what Hitler called the *Blitzkrieg*, the Gilera works team was in Genoa with Serafini ready to do battle with rival teams such as Moto Guzzi! And no sooner was the fighting over than race meetings were being run again, even though the rest of the country was virtually an empty shell.

Marte – Military Innovation

Like its British counterpart Norton, Gilera built world-class racers powered by overhead camshaft engines, but the machines they built for the army during the Second World War were far less glamorous, being powered almost exclusively by side-valve engines.

Gilera's main model for military duties when war was declared by Mussolini on 10 June 1940 was the LTE, based on the civilian L/LE models. The military LTE, with its 498.76cc (84 × 90mm), made its début in 1936 – and was to remain in production until 1944. It was to be used both in Italy and overseas and was employed in either solo form or with a sidecar attached, as service requirements demanded.

However, a much more serious attempt at a purpose-built military motorcycle was to result in the Marte (Mars). Although based on the LTE and its 500 side-valve engine, there were some important changes. For starters the engine was given more power, thanks to an alloy cylinder head, higher compression (but still low, at 5:1) and revised timing. Power output was 14bhp at 4,800rpm.

But if the engine was similar, the transmission certainly wasn't. The first change was the gear, which replaced a chain for the primary drive. And, although the gear change continued to be by

Marte – Military Innovation

hand, the drive was then taken via a pair of bevel gears to a shaft to transmit power to the rear wheel. The bevel box drove the rear wheel from the crown wheel and included a spur gear set that, in turn, drove a cross-shaft ahead of the rear axle. A second set of spur gears at the sidecar wheel (most Martes were built as three-wheelers) transferred the drive back again so the two rear wheels were in line. The sidecar gears also included a dog clutch controlled by a hand lever on the motorcycle to engage the drive as required.

Whilst the frame remained much as the LTE, the offside of the rear fork was totally redesigned and now became a bell-crank, manufactured using steel pressings welded together. The suspension medium remained as before – with girder front forks, and Gilera's patented horizontal boxed springs to provide movement at the rear.

The sidecar wheel was suspended by a trailing arm, with the 'chair' being mounted on the right so the drive from the rear wheel matche – this also being on the right.

This layout ensured that Gilera was the only motorcycle builder of the Second World War to offer a sidecar, with both the wheels of the motorcycle and sidecar having their own suspension systems, as well as offering drive to both wheels. Even BMW and Zündapp failed to match Gilera in this area, with only the sidecar wheel of their combinations being sprung – the rear wheel of the actual motorcycle remaining rigid.

Indeed, in the Marte, Gilera had designed and built a truly innovative machine, and one which was unique amongst its military counterparts during the Second World War.

Gilera's most important contribution to the Italian effort during the Second World War was the Marte 500 single. It used side-valves and shaft final drive.

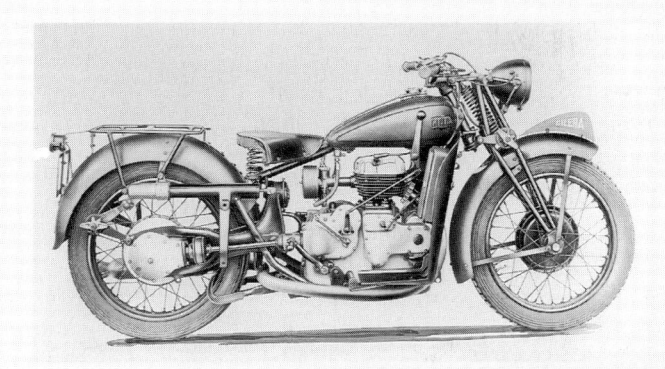

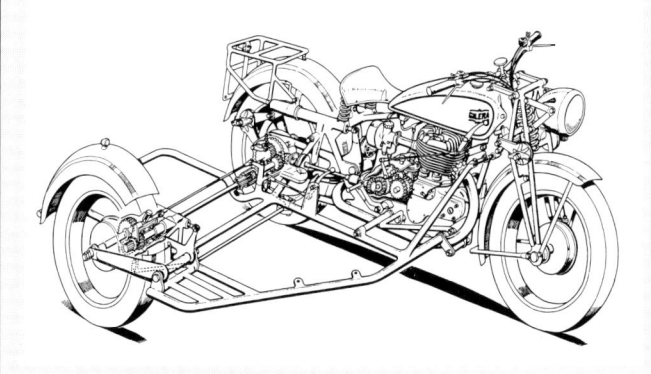

The sidecar version of the Marte was unique amongst wartime motorcycles as it provided drive for both the rear wheel of the motorcycle and the sidecar.

2 Saturno

Gilera first began work on overhead valve singles during the mid-1920s, mainly for racing use only. The most widely used was the VT Corsa (1924–1930) with a capacity of 490cc and long-stroke dimensions of 79 × 100mm.

THE TRE VALVOLE

The theme was picked up again with the Tre Valvole 500 which Mario Mellone designed in 1933. This featured two inlet valves and a single exhaust valve. Next came the Bitubo VT with twin exhaust ports the following year and, finally, the hugely successful VT single port series which ran from 1935 through to 1941. This series was then referred to as the Four Bolt and the Eight Bolt, and they dominated their class as fast touring bikes in the period just prior to World War II.

The Mario Mellone-designed Tre Valvole (Three-Valve) single of 1933, featured two exhaust and one inlet valve.

All shared the same engine size – 498.7cc (84 × 90mm). Other features included a power output ranging from 20 to 24bhp for the production models, although a higher output was available for the hand-built T Gran Sport which not only came with lumpier cams and higher compression pistons, but also a Dell'Orto SS30 racing carb. Maximum speed on this bike was a shade over 90mph (144km/h) and it dominated the 1939 Milano–Taranto; the outright winner being Ettore Villa.

THE SATURNO ARRIVES

This motorcycle can be considered as the true forerunner to the famous Saturno, which followed almost immediately. It was the work of Ing. Giuseppe Salmaggi who returned from Belgium where he had worked for Sarolea and who, together with FN, represented that country's two main motorcycle manufacturing outlets.

The Saturno gained its name from Gilera's astronomical theme which saw other models appear with names such as Nettuno (Neptune). Making its début in the spring of 1940 in competition trim, it won two races at the Italian Junior Championships at Palermo and Modena ridden by the diminutive Massimo Masserini, on both occasions beating a pack of Moto Guzzi Condors in the process.

In total it appears only five Saturno engines were built up to the end of 1945, even though the factory publicized the model both in Sport and Turismo trim.

Saturno

Designed in the late 1930s by Ing. Giuseppe Salmaggi and first shown in 1939, but not entering full production until 1946, the Gilera Saturno can fully justify claims that it was one of the truly great motorcycles produced by the Italian industry. Powered by an overhead-valve engine, the 500 Saturno Sport (seen here) produced 22bhp at 5,000rpm and was good for 85mph (136km/h). This is one of the original versions, built between 1946 and 1950.

	Saturno Turismo (1946–1959)
Engine:	Air-cooled, single-cylinder, four-stroke, ohv with cast-iron barrel and alloy head; (some early engines had iron heads)
Bore:	84mm
Stroke:	90mm
Displacement:	498.76cc
Compression ratio:	6:1
Maximum power (at crank):	18bhp @ 4,500rpm (Sport version, 22bhp @ 5,000rpm)
Lubrication:	Wet sump
Ignition:	Magneto, Marelli, MCR/40
Fuel system:	Dell'Orto RDF28 carburettor
Primary drive:	Gear
Final drive:	Chain
Gearbox:	Four-speed
Frame:	Duplex, steel cradle, single front downtube
Front suspension:	Early bikes girder forks; later telescopic type
Rear suspension:	Early bikes horizontal spring boxes/swinging fork; later bikes twin shocks with swinging arm
Front brake:	Drum. Early bikes single-sided; later bikes 220mm full-width
Rear brake:	Single-sided
Wheels:	19in front and rear
Tyres:	3.25 × 19 front and rear
Wheelbase:	58in (1,470mm)
Dry weight:	370lb (168kg)
Fuel tank capacity:	3imp. gal (14ltr)
Top speed:	75mph (120km/h)

PRODUCTION BEGINS

However, in reality the Saturno did not enter production until the 1946 model year: Sport, Turismo and Competition. The three versions were virtually identical. They differed only in the material used for the cylinder and head, weight of crankshaft flywheels, compression ratios, carburettor size, camshaft profiles, valve diameters, valve timing, type and size of tyres, and colours. These first bikes all used the traditional Gilera girder front fork assembly featuring a large central spring; whilst the rear end was taken care of by a suspension system patented by the company in 1935. This featured coil springs in horizontal sleeves with adjustable friction dampers.

THE ENGINE

Although much of Salmaggi's engine was new, the basic capacity of bore and stroke measurements remained from the VT series.

The Turismo and Sport (plus the Militaire which was built from 1948 until 1959), had a cast-iron cylinder, whilst that of the racer was in aluminium alloy with a pressed-in cast iron sleeve.

The valves were angled at 70 degrees and supported by enclosed hairpin springs. Some of the early production models were fitted with a cast-iron cylinder head, but most Saturnos were equipped with an alloy head. An aluminium piston had four rings and was connected to the steel connecting rod by a 0.8in (20mm) gudgeon pin; there was no small-end bush, the gudgeon pin bearing directly on the con-rod eye.

A built-up crankshaft was secured by means of tapers and fixing bolts. The

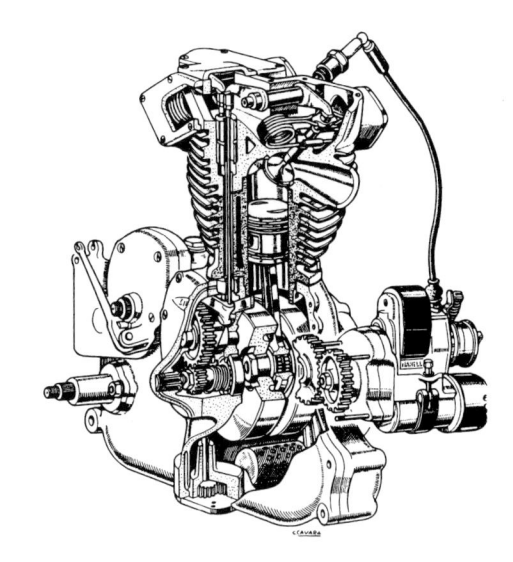

The 498.76cc (84 × 90mm) Saturno Sport engine showing details including: hairpin valve springs, four-ring piston, full-circle crankshaft flywheels, double rollers big-end, mag/dyno and kick-starter mechanism.

diameter of the pin that acted as a bearing surface for the cageless big-end rollers was 1.2in (30mm).

THE CLUTCH AND GEARBOX

The clutch was a bit of an odd-ball, being described in the factory workshop manual as 'dry', although it was in fact 'wet' because, although it did not operate in an oil bath, the five friction and four plain plates were 'lubricated' by a mist of oil during operation.

The four-speed foot-operated gearbox was driven by primary gears from the crankshaft. Ignition, from a Marelli clockwise-driven magneto, had manual advance, although later models with telescopic front forks came with automatic advance.

A gear-driven oil pump circulated lubricant from the wet sump contained at the

A Saturno Sport in action far away from its Arcore home, in Venezuela. This photograph dates from 1952 and shows local champion Emilio Varacca who set a new lap record for production machines at the San Monica circuit, at an average speed of 70mph (113kph).

Colourful 'Saturno Sport' front mudguard emblem – a feature of Gilera's series production motorcycles.

base of the crankshaft. On the roadsters a dynamo was locked at the front of the crankcase and was driven by a trio of gears in a horizontally arranged assembly, activated by the crankshaft. The forwardmost of these also drove the rev counter via a cable to the Veglia instrument on the racing models.

Saturno

A 1948 500 Saturno. This period photograph shows father and son with the family's motorcycle. Today the Saturno is one of the most sought after bikes in Italy, equal in status to the BSA Gold Star in Britain. Both machines were not only sports roadsters, but successful in various branches of bike sport.

From 1952 the Saturno roadster was built with telescopic forks and swinging arm rear suspension with twin vertical shocks.

MAJOR CHANGES

The major change to the Saturno range came at the end of 1950, with telescopic front forks, full-width alloy front brake hub and a different shape fuel tank, with cut-outs for the rider's knees and the capacity increased to 3gal (14ltr). A year later, and the rear suspension was changed to vertically mounted twin shock absorbers, with enclosed coil springs.

RACING MODELS

The more sporting models, and in particular the racing version, was prone to suffer from excess wear of camshafts, pushrods, rockers and valve springs – mainly, it must be said, due to the lubrication system not being improved over the standard low performance Turismo roadster.

Immediately after the war the factory provided a number of Competizione models

The 1948 version of the Saturno Sanremo racer. Introduced the previous year, this version had been built following Carlo Bandirola's outstanding victory on a Saturno Competizione at the recently opened Ospedaletti circuit at San Remo.

Grand Prix star Fergus Anderson tries the riding position of the Saturno on which Arcisco Artesiani had just won the 500cc category of the 1948 Italian Grand Prix at Faenza (a street circuit in Milan). The choice of this venue was made as the Monza Autodrome was still out of commission following extensive war damage.

to riders such as Bandirola, Clemencigh, Masserini, Ruggeri and Soprani.

Luigi Gilera, who still piloted a sidecar until he was fifty-years-old and, at the same time, managed the small race shop set up to deal exclusively in the single-cylinder models, built a special 584cc (84 × 105mm) Saturno engine for his own outfit.

THE SANREMO MODEL

During 1947, when some 13 Competiziones were built, Gilera had great success at the recently opened Ospedaletti circuit at San Remo. Bandirola, riding one of the factory entered Saturnos, won a hotly contested race. In recognition of his outstanding victory the single-cylinder racer was called the Sanremo, opening a golden era which was to run for half a decade. Arciso Artesiani won the 500cc class at the GP *Delle Nazoni* (Grand Prix of Nations) at the Faenza circuit in Milan that same year on another Saturno, overcoming the combined strength of both the Guzzi and Norton teams.

Saturno

Saturno Sanremo (1947–1951)

Engine:	Air-cooled, four-stroke, ohv, single-cylinder
Bore:	84mm
Stroke:	90mm
Displacement:	498.76cc
Compression ratio:	6.5:1
Maximum power (at crank):	35bhp @ 6,000rpm
Lubrication:	Wet sump
Ignition:	Magneto, Marelli
Fuel system:	Dell'Orto SS35 carb
Primary drive:	Gear
Final drive:	Chain
Gearbox:	Four-speed, close ratio
Frame:	Steel, tubular. Engine as stressed member. Single front downtube
Front suspension:	Blade-type, with central spring
Rear suspension:	Horizontal boxed type, with friction adjusters
Front brake:	Drum. Single leading shoe
Rear brake:	Drum. Single leading shoe
Wheels:	Front 21in; Rear 20in
Tyres:	Front 3.00 × 21; Rear 3.25 × 20
Wheelbase:	55.5in (1,410mm)
Dry weight:	264.5lb (120kg)
Fuel tank capacity:	3imp. gal (14ltr)
Top speed:	112mph (180km/h)

Note:
A larger 582cc (84 × 105mm) engine was also constructed for sidecar use.

A major difference between the Competizione and Sanremo customer racers was the fitment of the Dell'Orto 32 and 35 SS carburettors respectively. Also the Sanremo benefitted from a higher 6.5:1 compression ratio (5.8:1) and more power 35 (32)bhp. The Sanremo also boasted blade girder forks, a full-width front brake and pressed steel rear mudguard which doubled as a racing plate.

Saturno Sanremo racing engine details: three-ring piston, hairpin valve springs, timing gears and magneto ignition.

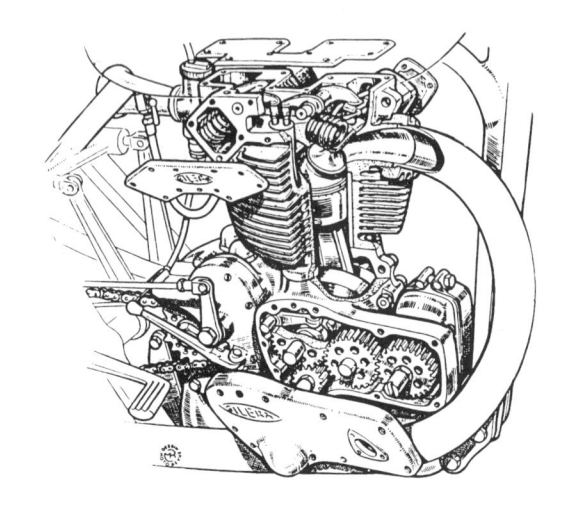

Saturno

A 1949 Sanremo racer in action. Note blade-type front forks and full-width aluminium brake hub.

The production Saturno Corsa (racing) was sold between 1951 and 1957. The final version (shown) was first offered in 1956 and was a most handsome and purposeful machine. Its specification included telescopic front forks, swinging arm rear suspension and the full-width front brake.

The Sanremo title was apt, as first Bandirola repeated his performance with another victory in 1948, followed by Masetti (1949), Colnago (1950) and Valdinoci (1951).

Forty-nine examples of the Sanremo were built in 1949 and were not only raced with great success in Italy by up-and-coming riders such as Masetti, Alfredo Milano, Liberati, Colnago and Valdinoci, but also by riders in France, Hungary, Spain, Switzerland and Belgium. One was even converted to a double overhead camshaft operation by the Swiss sidecar driver Keller.

THE CORSA

In 1951 the Sanremo was updated to become the Corsa. In an attempt to provide better cooling, more comprehensive finning was introduced for the head and cylinder whilst the oil sump capacity was increased. The compression ratio was raised to 7.06:1 and other improvements saw the power rise to 38bhp at 6,000rpm.

Additionally, Alfredo Milani almost scored a sensational victory in the first round of the World Championship, staged at Montjuich Park, Barcelona, on one of the new bikes. With telescopic forks fitted

Saturno Corsa (1951–1957)

Engine:	Air-cooled, four-stroke, ohv, single-cylinder
Bore:	84mm
Stroke:	90mm
Displacement:	498.76cc
Compression ratio:	7.06:1
Maximum power (at crank):	38bhp @ 6,000rpm
Lubrication:	Wet sump
Ignition:	Magneto, Marelli ST6148
Fuel system:	Dell'Orto SSF 35M carb
Primary drive:	Gear
Final drive:	Chain
Gearbox:	Four-speed, close ratio
Frame:	Steel, tubular. Engine as stressed member. Single front downtube
Front suspension:	Telescopic fork with enclosed springs and stanchions
Rear suspension:	1951 Horizontal boxed type, with friction adjusters; 1952 Twin shock, swinging arm
Front brake:	220mm drum, full width aluminium hub, SLS
Rear brake:	185mm drum. Single-sided, steel hub, SLS
Wheels:	19in front and rear
Tyres:	Front 3.00 × 19; Rear 3.25 × 19
Wheelbase:	54in (1,380mm)
Dry weight:	275.5lb (125kg)
Fuel tank capacity:	4.18imp. gal (19ltr)
Top speed:	1951 118mph (190km/h); 1956 130mph (210km/h)

Note:
The following gearing ratios (sprockets) were available: 17 × 44, 17 × 46, 18 × 46, 18 × 44, 19 × 46, and 19 × 44.

Spanish Grand Prix, Montjuich Park Barcelona, 4 October 1953. Ernesto Soprani (29) on his Saturno Corsa leads a four-cylinder Gilera (34) and a Norton single (38).

he had built up a large lead over no less that eight, four-cylinder machines from Gilera and MV Agusta, plus the British Nortons; however, with only a few laps left he was forced out with broken valve gear.

TELESCOPIC FRONT FORK

For 1952 Gilera adapted the telescopic fork and hydraulically operated rear shock absorbers as standard on the Corsa. The forks were made by the Arcore factory itself. Although its official name was Corsa, it was more generally known as the *Piuma* (feather) because of the improved handling and ease of control.

Nine were built in 1952, fourteen in 1953 and a small number in 1954. Then, in 1955, Gilera returned to racing the Saturnos, but only in events such as the Milano–Taranto. They were rewarded when Domenico Fenocchio won the Sports class of the event that year.

Gilera Saturno Corsa engine, note rev-counter drive, gearbox and remove float for Dell'Orto SS carburettor.

Saturno

TWIN-CAM ENGINE

Besides the pushrod model there was also the Franco Passoni-designed twin-cam engine; this was essentially a Corsa model with a significantly refined engine. This produced 45bhp at 8,000rpm and was developed in 1952 and 1953.

One of the two solo machines built was used by the English rider Harry Voice (a close friend of Geoff Duke) in the 1953 Manx Grand Prix. After running around fifteenth position near the end of the race Voice was forced out as the result of an oil leak. This dohc model had made its race début in April 1953, and its sole victory came in the non-championship French Bordeaux GP ridden by veteran Georges Monneret.

Nearside of the Corsa engine with magneto, clutch cover and oil filler and drain plugs all visible.

Lovely action photograph of Saturno Corsa rider Giovanni Pio. His Saturno was one of the very last constructed and displays the attractive lines of the machine to perfection. The year is 1957.

Saturno

> **Saturno Bialbero (Twin cam) (1952–1953)**
>
> | | |
> |---|---|
> | Engine: | Air-cooled, four-stroke, ohv, single-cylinder |
> | Bore: | 84mm |
> | Stroke: | 90mm |
> | Displacement: | 498.76cc |
> | Compression ratio: | 8:1 |
> | Maximum power (at crank): | 45bhp @ 8,000rpm |
> | Lubrication: | Wet sump |
> | Ignition: | Magneto, Lucas |
> | Fuel system: | Dell'Orto SS 38mm carb |
> | Primary drive: | Gear |
> | Final drive: | Chain |
> | Gearbox: | Four-speeds, close ratio |
> | Frame: | Steel, tubular. Engine as stressed member. Single front downtube |
> | Front suspension: | Telescopic fork with enclosed springs and stanchions |
> | Rear suspension: | Twin shock, swinging arm |
> | Front brake: | 220mm drum, full-width aluminium hub, 2LS |
> | Rear brake: | 185mm drum, single-sided, steel hub, SLS |
> | Wheels: | 19in front and rear |
> | Tyres: | Front 3.00 × 19; Rear 3.25 × 19 |
> | Wheelbase: | 54in (1,380mm) |
> | Dry weight: | 274lb (125kg) |
> | Fuel tank capacity: | 4.18imp. gal (19ltr) |
> | Top speed: | 124mph (200km/h) |

The first public appearance of the dohc Saturno Bialbero Corsa (double-camshaft racing) came during practice for the 1952 Italian GP at Monza; it also made an appearance during the Spanish GP a month later. Development was scrapped due to the success of the four-cylinder models.

Saturno

The Saturno was also built as a motocross machine. The final batch of the Saturno Cross model was constructed during 1956. The police also used a number of these machines for its display team during the late 1950s and early 1960s. Here, one such bike leaps twelve team members during a display.

The ultimate Saturno Super Sport roadster of 1956–1957; spoilt only by the over-flashy twin-barrel exhaust designed by the Abarth company, who were more well known for their tuning abilities with Fiat cars.

> **Saturno Cross (1952–1956)**
>
> | | |
> |---|---|
> | Engine: | Air-cooled, four-stroke, ohv, single-cylinder |
> | Bore: | 84mm |
> | Stroke: | 90mm |
> | Displacement: | 498.76cc |
> | Compression ratio: | 7.06:1 |
> | Maximum power (at crank): | 38bhp @ 6,000rpm |
> | Lubrication: | Wet sump |
> | Ignition: | Magneto |
> | Fuel system: | Dell'Orto SSF 38M carb |
> | Primary drive: | Gear |
> | Final drive: | Chain |
> | Gearbox: | Four-speed |
> | Frame: | Full cradle steel, with single front downtube |
> | Front suspension: | Enclosed stanchion, telescopic |
> | Rear suspension: | Twin shock, with enclosed springs, steel swinging arm |
> | Front brake: | Drum, full-width aluminium hub, 2LS |
> | Rear brake: | 185mm drum, single-sided, steel hub |
> | Wheels: | Front 21in; Rear 18in |
> | Tyres: | Front 3.00 × 20; Rear 4.00 × 18 |
> | Wheelbase: | 55in (1,400mm) |
> | Dry weight: | 275.5lb (125kg) |
> | Fuel tank capacity: | 2.2imp. gal (10ltr) |
> | Top speed: | 82mph (130km/h) |

MOTOCROSS

Another facet of the Saturno was motocross, a number being constructed between 1952 and 1956. In fact the winner of the 1955 Milano–Taranto, Domenico Fenocchio, was the factory's main off-road star, winning three consecutive Italian Motocross Championships from 1953 to 1955.

Its engine, at least, was to the same specification as the Corsa racer; it was also amazingly like the racer, except for a straight pipe exhaust, a 21in (533mm) front rim, sump bash plate, smaller brake hubs, modified suspension (including slightly forward mounted front-wheel axle forks), new seat, tank and mudguard and knobbly tyres.

ROADSTER DEVELOPMENT

As for the roadster, this continued in production from 1952 onwards featuring a full-width alloy brake hub at the front, telescopic forks and swinging arm suspension with twin shocks.

THE NETTUNO 250

Besides the Saturno there was also the smaller Nettuno 250 – 246.95cc (68 × 68mm). Produced in two versions, the Turismo and Sport, it was manufactured between 1946 and 1954. Cycle parts development largely mirrored that of the Saturno – for example girders, blade girders and finally teles. A special version was built for use in long-distance trials

Saturno

Unlike its larger Saturno brother, the smaller Nettuno (Neptune) featured 'square' 68 × 68mm bore and stroke dimensions, giving a capacity of 246.95cc. In Turismo guise it turned out 11bhp at 5,200rpm, whilst the more highly tuned Sport (shown here) gave 14bhp at 6,000rpm.

Like the Saturno, the magneto/dynamo of the Nettuno was mounted at the front of the semi-unit construction crankcase assembly. Lubrication was of the wet sump variety which accounted for the large area of the base of the crankcase; production ceased during 1954.

Saturno

	Nettuno 250 (1946–1954)
Engine:	Air-cooled, single-cylinder, ohv, with cast-iron barrel and alloy head (some early engines had iron heads)
Bore:	68mm
Stroke:	68mm
Displacement:	246.95cc
Compression ratio:	Iron head 6:1; alloy head 6.8:1
Maximum power (at crank):	Iron head 11bhp @ 5,200rpm; alloy head 14bph @ 6,000rpm
Lubrication:	Wet sump
Ignition:	Magneto, Marelli MCR4D
Fuel system:	Dell'Orto RCF25 carburetter
Primary drive:	Gear
Final drive:	Chain
Gearbox:	Four-speed
Frame:	Tubular, steel
Front suspension:	Early bikes girder forks; later telescopic type
Rear suspension:	Early bikes horizontal spring boxes/swinging fork; later bikes twin shocks with swinging arm
Front brake:	Drum. Early bikes single-sided; later bikes aluminium full-width
Rear brake:	Single-sided
Wheels:	Front 2.75 × 19; Rear 3.00 × 19
Tyres:	Front 3.00 × 20; Rear 4.00 × 18
Wheelbase:	55.5in (1,410mm)
Dry weight:	308.5lb (140kg)
Fuel tank capacity:	2.85imp. gal (13ltr)
Top speed:	Iron head 64mph (103km/h); alloy head 78mph (125km/h)

Stateside Gilera enthusiast Todd Fell with his Saturno Sport, New Orleans, Louisiana.

Saturno

Former Gilera works rider and the 1957 500cc World Champion, Libero Liberati, pictured with his 'private' Saturno Corsa model at the Vallelunga circuit near Rome, October 1961. Early the following year, whilst riding the same machine, he was fatally injured in a road accident near his Terni home.

(*see* Chapter 7). But the smaller single was never as popular as the half-litre bike.

THE LAST SATURNOS ARE BUILT

The final two Saturno Sport roadsters were sold in 1960, although they had been built the previous year. In total around 6,450 examples of all the Saturnos were manufactured, of which 170 were racers.

LIBERO LIBERATI

The last great Saturno feat came in 1961 when the 1957 500cc World Champion, Libero Liberati, was only just pipped for the Italian Senior title by Ernesto Brambilla on a works Bianchi twin.

Liberati's bike featured one of the twin leading shoe brakes from a 1957 model four cylinder and an engine breathed on by the factory's ex-race staff. Rumour had it that it even had a new lubrication system.

Sadly, Liberati met his death near his home town of Terni whilst trying out his beloved Saturno on public roads in readiness for the season's racing, which was due to start at Modena a week later. Thirty-six-years-old at the time of the accident, Liberati had started racing soon after the war, before being signed up by Gilera for the 1952 season. In 1955 and 1956 he won the 500cc Senior Italian Championships riding four-cylinder models. In winning the following year's 500cc World title, he finished first in four of the six rounds – Germany, Belgium, Ulster and Italy.

After Gilera quit the GP scene at the end of 1957, Liberati continued racing his private Saturno, but with some 'back-room' help from factory staff. Perhaps only a rider of Liberati's ability would have kept what was, in reality, a machine from an earlier era, right up at the front for so long. This achievement above all else is a fitting tribute to one of Italy's all-time greats.

3 Grand Prix Glory

AIR-COOLED, SUPERCHARGED 250 FOUR

Just prior to Italy's entry into the Second World War, the Gilera racing department employed the experience accrued with the 500cc machine as the basis for the development of an air-cooled 250cc four, which had its supercharger mounted in front of the crankcase instead of behind the cylinders. Although never actually raced, this smaller engine was to prove of great significance in the development of the post-war 500cc (and 350cc) Gilera four.

250cc Four-Cylinder Projects

During 1939, Giuseppe Gilera asked Piero Taruffi to recommend a designer capable of creating a brand-new multi-cylinder job. Taruffi, perhaps naturally, offered the name of his former OPRA/Rondine associate, Ing. Piero Remor.

The machine was to be a 250, to complement the Rondine-inspired 500 Quattro. And again, perhaps naturally, Remor created a four, but in miniature. And it had a whole host of competition from the likes of DKW, Benelli and Moto Guzzi. Of the opposition, only Benelli were to build a four – a water-cooled dohc equipped with a Cozetto blower.

But Remor viewed the German DKWs as his major threat, not only did he feel they were exceedingly quick, but had a mega-budget with which to fund additional development.

Remor started by electing to use not only supercharging, but also central gear drive to the double overhead camshafts. But these features apart, Remor's quarter-litre Gilera four owed little to his earlier creations. For a start the blower was constructed at the front of the heavily-finned crankcase assembly, with the exhausts facing rearwards. And, unlike the Benelli design, Remor opted for air-, not water-cooling. He inclined the cylinders 30 degrees from the vertical, and specified a wet-sump lubrication system.

Development spread into 1940 and, although the engine was completed and test run, it was not fitted into a chassis before Mussolini declared war on the 10 June that year. Even though the unit survived the conflict, it didn't survive the FIM's post-war ban on supercharging and so it languished throughout the late 1940s, even though at least one entry was made for it during the 1945 season. However, although it was destined never to turn a wheel in anger, certain aspects of the design were to surface in the newly created Remor 500 four which made its début in 1948 – and was to be the forerunner of the World Championship winning machine of the 1950s.

Almost a decade later in 1956, the 250 Gilera four was reborn under Remor's successor, Franco Passoni. Essentially this new machine employed a couple of the 125 twin-cylinder GP engines in a common crankcase that was to be encased within a streamlined shell. The design was finished in September 1956, but with the sudden death of its main supporter, Giuseppe Gilera's only son, Ferruccio, only a few short weeks later from a heart attack, the 250/4 project was killed off, too. In fact the stillborn 1956 250 four could be truthfully labelled as Gilera's final 1950s Grand Prix project. Had Ferruccio lived, his dream of Gilera competing in all four solo capacity classes (125, 250, 350 and 500cc) would have been realized.

Grand Prix Glory

> **250cc Four-Cylinder Projects** *(continued)*
>
> As for that final Gilera 250 four's creator, Passoni, he was to leave Gilera during the early 1960s, but as a condition of his being allowed to leave he had to sign an agreement that he would not work for another motorcycle company and thus share Gilera's secrets (Giuseppe Gilera had not forgotten the Remor defection to MV Agusta!). And thus Franco Passoni began a new career in the burgeoning computer industry.
>
> *The experimental 248cc (52 × 58mm) aircooled dohc supercharged (using a gear/driven vane-type blower) four. Built in 1940, the war halted its development.*

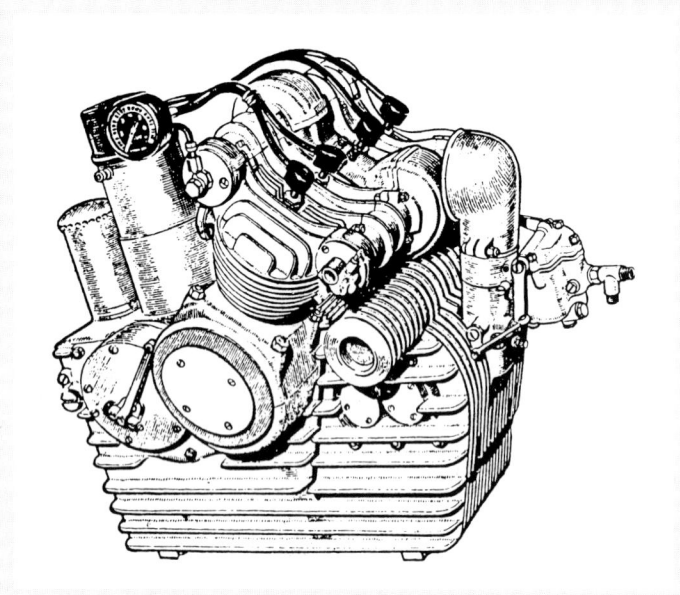

FIM BAN SUPERCHARGING

Immediately after the war, the Arcore factory dug out examples of the old water-cooled Rondine-based four but, although some respectable results were achieved in both solo and sidecar categories, the removal of the supercharger to conform with the FIM's newly introduced ban on blowers saw the power drop by no less than *half* – from 90 to 45bhp! With virtually no saving in weight, the machine was no longer the great force it had once been, so an early decision was taken to build an entirely new version that would incorporate the design advances pioneered in the 1940 250.

REMOR'S REDESIGN

The new machine was the work of Ing. Piero Remor and was ready by the end of 1947. Its air-cooled engine, with the cylinders inclined thirty degrees forward from the vertical, pumped out 55bhp at 8,500rpm. The four-speed gearbox was constructed in unit with the engine and the assembly mounted in a pressed steel chassis, which had blade girder-type front forks with a central coil spring and torsion-bar swinging arm rear suspension. The large diameter alloy brake hubs were mounted in 20in alloy rims. Unlike the pre-war 500, the new model sported wet-sump lubrication. Carburation was looked

The first post-war 500cc-class Gilera four owed its existence to pre-war design, being simply a normally aspirated version of the old bike. Then in 1948 a new four arrived. The work of Ing. Piero Remor, this was to suffer serious lubrication gremlins which were not to be fully resolved until Remor had quit Gilera for rivals MV Agusta.

The 1948 496.692cc (52 ×58mm) four-cylinder air-cooled engine.

after by the fitment of a new type of Weber instrument, one of which each fed a pair of cylinders. Early ignition problems were solved by the introduction of a Marelli magneto, mounted vertically behind the cylinder block. A wet multi-plate clutch was used, while the drive to the primary gears was taken from between the first and second cylinders.

Remor spent much time selecting the best and also the lightest materials; this was to pay off handsomely with the new bike weighing in at a surprisingly low 275lb (125kg), making it one of the lightest in its class, and this was at a time when its rivals were mainly single-cylinder types.

RACE TESTING BEGINS

Although completed during 1947, the machine was not wheeled out in anger until the following year. Its initial testing was carried out by factory rider Carlo

Bandirola over the Milano–Bergamo *autostrada*; whilst its race début came on 9 May at Cesena, in the experienced hands of Nello Pagani.

But a shock was in store as, quite bluntly, Pagani labelled the new bike 'unrideable'. This was not the news that Remor, or for that matter his boss Giuseppe Gilera, wanted to hear, but it was nonetheless true.

So, although Massimo Masserini gave the Remor-designed four its first victory in July, much of 1948 was spent in attempting to put right the machine's failings. These ranged from poor handling to mechanical problems, centred on the lubrication system. Just how low Pagani rated the 1948 four is illustrated by the fact that he opted to race his own 'private' single-cylinder Saturno instead …

Even so, Massimo Masserini swears to this day that the bike was not anything like as bad to ride as Pagani made out. His theory is backed up by the fact that he recorded the new four's first ever GP victory, the Italian, over the 3-mile (4.8km) Milanese Faenza street circuit in September 1948. But Masserini did concur with the engine's unreliability at this stage, and was later to reveal that Ing. Remor would not (or could not) accept faults with his design. It is also interesting to note that the lubrication problem was solved by his assistant Alessandro Colombo.

Four-cylinder 500GP (1948–1951)

Engine:	Air-cooled, four-stroke, dohc, across-the-frame four-cylinder
Bore:	52mm
Stroke:	58mm
Displacement:	496.692cc
Compression ratio:	10:1
Maximum power (at crank):	1948 48bhp @ 8,500rpm; 1951 50bhp @ 9,100rpm
Lubrication:	Wet sump
Ignition:	Magneto, either Vertex Scintilla or Marelli
Fuel system:	4 × carburettor, either Weber 28mm or Dell'Orto SS 29 or 30mm
Primary drive:	Gear
Final drive:	Chain
Gearbox:	Four-speed
Frame:	Duplex, steel cradle
Front suspension:	1948 Blade fork with central spring; 1951 Telescopic fork, fully enclosed
Rear suspension:	1948 Spring boxes and friction adjusters; 1951 Twin shock, enclosed type
Front brake:	1948 Drum, single-sided, aluminium; 1951 Drum, full-width, aluminium
Rear brake:	Drum, single-sided
Wheels:	20in front and rear
Tyres:	Front 3.00 × 20; Rear 3.25 × 20
Wheelbase:	58.5in (1,490mm)
Dry weight:	1948 275.5lb (125kg)
Fuel tank capacity:	4.4imp. gal (20ltr)
Top speed:	1948 125mph (200km/h); 1951 128mph (205km/h)

Protar Gilera 500cc Four-Cylinder Racing Motorcycle 1/9th Scale Kit

This machine guaranteed its place in history when Bob McIntyre lapped the Isle of Man Mountain Circuit in the 1957 Jubilee TT at over 100mph (160km/h). The fully streamlined Gilera was virtually unbeatable that year and Libero Liberati took the 500cc World Championship title. It was a sad day for racing when the Gilera factory pulled out of competition at the end of that season.

The model depicted here captures the big Ancore four perfectly in miniature. From the finning on the cylinders, to the float chambers for the carburettors, nothing has been overlooked. There are springs included for the working suspension units and when the rear wheel is turned the chain operates the drive and the gearbox. The kit itself is simplicity to build and, of course, the longer one spends painting it, the better the finished article looks. This machine is an absolute must for any serious motorcycle model builder as it represents one of the finest racing motorcycles ever created.

Ian Welsh

THE 1949 CHAMPIONSHIP SERIES

1949 was the first year of the World Championship Series and Masserini had retired over the winter (being forced back into running the family business in Bergamo, following the death of his mother). Officially the Gilera team comprised Pagani, Bandirola and newcomer Arciso Artesiani. However, in truth, Pagani had deepened his rift with Remor and spent the year on his own Saturno. Pagani was to prove a source of embarrassment for Remor, if not Gilera, as, after a superb fourth in the Swiss GP on his own Saturno, he was asked to 'please try the four again'. He responded by winning the Dutch TT at Assen the following weekend, beating champion elect, Les Graham (AJS twin). But Pagani could do no better than fifth in the next round in Belgium, slowed down, it was determined later, by a valve seat coming loose early in the race, causing a loss of power.

In the Ulster GP a lack of course knowledge also slowed his efforts, but he still came home third. Then, in the final round on home soil he led team-mate Artesiani to record a memorable Gilera one-two.

This promoted Pagani to runner-up in the World Series. Actually, with Graham crashing out in the final round, the decision not to race at the inaugural round in the Isle of Man by the Gilera squad, coupled with Pagani riding the single rather than the four in Switzerland, virtually handed AJS and Graham the Championship.

REMOR QUITS FOR MV

Then came controversy, with first Remor and Artesiani quitting Arcore to join MV Agusta, followed by chief mechanic Arturo Magni.

Grand Prix Glory

TARUFFI RETURNS

But the Gilera squad rallied, with Taruffi being re-engaged, this time as team manager. Giuseppe Gilera also promoted Remor's former assistants, Alessandro Columbo and Franco Passoni, to joint heads of the technical department.

The revised design team was obviously pressed for time to carry out any major changes for the start of the fast-approaching 1950 season (and were not

(Left) *The Swiss Grand Prix at Berne, 3 July 1949. A factory mechanic with Arcisco Artesiani's 500 four-cylinder Gilera which finished runner-up that day to Les Graham's AJS Porcupine.*

(Above) *Carlo Bandirola with the 1949 four-cylinder Gilera. This differed from the following year's model in the areas of brakes and suspension.*

(Left) *Nello Pagani on his way to victory in the Grand Prix des Nations at Monza in September 1949. Pagani was placed second in the overall 1949 500cc championship ratings behind Les Graham's AJS, with fellow Gilera team-mate Arcisco Artesiani third.*

helped by the fact that Remor had removed drawings for the next stage of development when he left for MV). So Gilera was forced to restrict its efforts to the cylinder heads (there were now two, instead of a single casting), the carburettor size (which was increased to 30mm with power rising to 52bhp at 9,000rpm) and, perhaps of most importance, the lubrication system was redesigned.

Another alteration, following Remor's defection, was that the rear springing reverted to the pre-war system, which consisted of horizontal cylindrical spring boxes at the base of the saddle, with friction damping. The braking system was also improved by the fitment of a full-width front hub, but the blade girder forks were retained, even though the rival British factories were by now using the more modern telescopics.

As in earlier years, the Gilera riders, now with Umberto Masetti as a replacement for the departed Artesiani, gave the Isle of Man a miss. Then came two victories for Masetti in Holland and Belgium, with Pagani as runner-up on each occasion (although it should be remembered that both the AJS and Norton works teams were forced out through tyre glitches).

GILERA WINS THE 1950 500cc WORLD CHAMPIONSHIP

Although Masetti could only manage a fifth in Ulster, he stormed back to finish a close second behind Geoff Duke (Norton) in the last round at Monza, and so in the process became the 1950 500cc World Champion.

But even though it had won the title, Gilera realized it had been extremely fortunate – the margin being a single point. If Duke had not encountered problems mid-season, the trophy would have been heading for Bracebridge Street, Birmingham, the Norton factory's headquarters.

As a point of interest, why had the aging single-cylinder Norton been in contention anyway? The answer lay in its excellent Featherbed frame and Roadholder telescopic front forks, not its outright speed.

TECHNICAL CHANGES

Ing. Passoni was charged with a major revision of the four-cylinder Gilera during the closed season of 1950–1951. Prompted by Norton's success he chose to use a round-tubed duplex chassis with swinging-arm rear suspension and hydraulically damped front forks similar to the Norton type. A switch was also made to smaller diameter rims (19in) and tyres.

Taruffi had, in the meantime, been busy securing the services of Pagani, Masetti and Alfredo Milani; at this time foreign riders were not being considered.

For 1950, improvements included larger carburettors, more power (52bhp at 9,000rpm), better lubrication, rear springing as used pre-war, and an up-rated, full-width front brake, but the blade-type forks were retained.

> **The British Influence**
>
> Whilst it is entirely true that Italy had almost always been a dominant force in the lightweight (125 and 250cc) categories (at least until the Japanese arrived) it certainly was not the case in the larger classes (350 and 500cc), where British machinery invariably ruled the roost.
>
> For a short period just prior to the outbreak of the Second World War both the German BMW twin and Gilera's Rondine-based four – with the aid of supercharging – took over. But the FIM's post-war ban on 'blowers' effectively put the British machines back in the frame.
>
> Except for Moto Guzzi and Benelli, Italian manufacturers knew little or nothing of the Isle of Man and its TT course before the 1939–1945 war, in fact, opinion was that they could still win without racing at the TT.
>
> Consequently, when the World Championship series was introduced in 1949, this thinking prevailed. But Gilera (and MV) soon discovered that they could not win as easily as they thought. They found that they needed far more than simply a powerful engine.
>
> The four-cylinder Gilera of 1950 was faster than the single-cylinder Norton, but the British machine was so much better as an overall package, due to its superb Featherbed frame and Roadholder forks.
>
> At first Gilera thought that all that was needed was a good bike, however, as even the Italians realized, at that time the British riders (including the Commonwealth) were better. This was largely believed to be due to their experience of riding in the TT.
>
> After Geoff Duke joined Gilera (at the beginning of 1953) it was he, together with the Irishman Reg Armstrong, who convinced Gilera that it was essential to race at the TT to fully develop their four-cylinder models.
>
> If you study photographs of the 1949/50 Gilera and the 1955 model, you can instantly see the British influence. Today, almost half a century later, it may seem strange as the top riders no longer grace the TT, but when Duke joined Gilera, one of the main things he told them was that he would not sign unless they raced at the TT!
>
> When the great riders are being discussed it is perhaps not Geoff Duke's name that heads the list, but in many ways he was truly great. No other rider before or since has been as smooth; he was also easy on the machine. As Regina chain engineer Ercole Villa once explained: 'Geoff Duke was the best. The chain on his Gilera was always perfect and would have lasted for five TT races. But at Monza where it is much easier for the machines, Alfredo Milani [another factory Gilera rider] would use ten chains to Duke's one.'

Other technical improvements were also introduced, the most notable being four instead of two carburettors (ranging in size from 25 to 28mm depending upon the circuit), higher lift camshafts and improvements in various materials. All this added up to an additional 2bhp.

DUKE/NORTON ARE SUPREME

However, in the end all this counted for little as the masterful riding style of Geoff Duke secured a famous double, with the Englishman taking both the 350 and 500cc titles – the first man in history to achieve this in a single season. The series – staged over eight rounds – saw Gilera score three victories: Spain (Masetti), France and Italy (Milani). The final round on home soil at Monza saw an impressive one-two-three for the Arcore team (Milani, Masetti and Pagani); it was also a warning to other teams of future intentions.

For 1951 Ing. Franco Passoni (who had replaced Remor) followed the successful Norton innovation of the previous year by adopting telescopic forks, a duplex frame and swinging arm rear suspension. Here, at the Spanish GP that year, Nello Pagani is watched by team manager Piero Taruffi.

Staged over the twists and turns of Montjuich Park, Barcelona in early April 1951, the inaugural Spanish GP attracted two four-cylinder Gilera entries, Pagani, plus Gilera's 1950 500cc World Champion, Umberto Masetti.

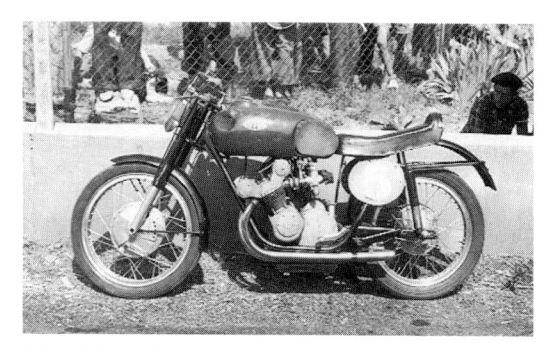

The Gilera four as it was in 1951. This photograph, taken at Albi the scene of the French Grand Prix that year, shows the various details of the machine, including the new suspension, frame and brakes.

1952 TWO AND THREE WHEELS

During 1952 Gilera made a massive effort on both two and three wheels. It should be remembered that they were the only large Italian factory to race sidecars during this era and much of this was thanks to Giuseppe Gilera's brother Luigi, himself a famous sidecar competitor. But Gilera's quest for three-wheel honours was dealt a massive blow when its leading contender Ercole Frigerio was killed during the Swiss Grand Prix. It was then left to Ernesto Merlo and Albino Milani (brother of Alfredo) to fly the Gilera flag in the 'chair' events. But Frigerio's record of three consecutive runner-up positions (1949, 1950 and 1951) was never matched by another Gilera sidecar driver; Milani's best being third in 1951 and Merlo fourth in 1952. Gilera withdrew from the class at the end of the year.

In the solo category a nose fairing had been tested at Monza back in September 1951; this being incorporated into the 1952-type fours. Other changes for that year were mainly restricted to a re-styling exercise. But, after Duke was sidelined

Grand Prix Glory

following an accident at a non-Championship event in Germany, Umberto Masetti took his and Gilera's second World title, with wins in Holland and Belgium to add to runner-up positions in Spain and Italy.

DUKE AND ARMSTRONG SIGN

Major changes occurred in 1953. For a start, Taruffi signed up Gilera's first-ever foreign riders with Ireland's Reg

(Left) For the 1954 Dutch TT, Gilera contracted the local rider H. Drikus Veer to ride one of the works fours. The Dutchman is seen here at the event in the Assen paddock that year. He came home in 8th position in the 500cc race.

(Below) A nose fairing was tested at Monza in September 1951 (where this photograph was taken). It was then incorporated into the 1952 fours, whilst the tank was re-styled and some other smaller modifications carried out.

Grand Prix Glory

(Above) *Gilera new boy Libero Liberati receiving attention at his pit during the Italian GP in September 1951. He ultimately completed the race in seventh position. The other members of the team – Milani, Masetti and Pagani finishing first, second and third, respectively.*

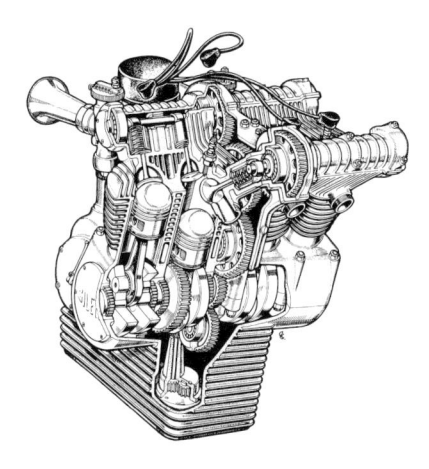

Cutaway view of the Gilera four-cylinder engine after its mini-redesign by Ing. Passoni, which included improved power output and increased reliability.

(Below) *The 1952-type four that Umberto Masetti rode at the Swiss Grand Prix on 18 May that year.*

(Above) *At the beginning of 1953, first Reg Armstrong and then Geoff Duke signed for Gilera. Both were formerly with Norton and they brought not only their riding talent, but also their experiences, which were to prove vital for the future development of the four-cylinder series. Amongst the features tried that year was a smaller 17in rear wheel. This was something Norton had been testing ... the tank seat and fly screen also showed British influence.*

Grand Prix Glory

Armstrong and England's Dickie Dale, to be followed later by the big prize, Geoff Duke. With Frenchman Pierre Monneret, there were four non-Italians. Add in Masetti, Milano and Giuseppe Colnago and Gilera had its strongest squad yet.

There followed an intensive testing programme, much of it at Monza, where Ing. Passoni incorporated a number of changes suggested by the foreign riders, notably Duke. These resulted in the so-called 'Nortanized' model, which, as the name implied, had a similar appearance to the Norton Featherbed. However, although there were many changes to the cycle parts, the engine specification remained largely unchanged.

THE ISLE OF MAN TT

For the first time, Gilera contested the Isle of Man TT with Duke, Armstrong, Dale and Milani. But after Duke had broken the lap record no less than *three* times during the race, he fell at Quarter Bridge at the start of his fourth lap, putting himself out of the race. The event had been billed as the 'Race of the Century', with the Norton and AJS teams lining up against the four-cylinder Gilera and MV's and Walter Zeller's lone factory BMW twin. Unfortunately, the race also claimed the life of MV team leader Les Graham.

DUKE BECOMES CHAMPION

After the TT Gilera riders dominated the series, except in Ulster, Duke taking the title, with Armstrong runner-up and Milani fourth. What happened to Masetti? Well, feeling that he had been demoted since the arrival of the foreign riders, and in particular Geoff Duke, and following a dispute with team manager Taruffi, he quit the team mid-season.

Although it had been an excellent season for Arcore, Gilera realized that this situation was unlikely to continue without further technical development; Guzzi, Norton and MV were all trying to close the gap.

MORE TECHNICAL DEVELOPMENT

By this time Ing. Columbo had left and race development was solely the responsibility of Ing. Passoni; it was he who instigated a major revision of the design for 1954.

The majority of the work was centred on the engine. The stroke was increased from 58 to 58.8mm giving a new capacity of 499.504cc. The sump was modified to allow the engine to be located lower in the frame, yet at the same time ground clearance was increased by tucking the exhaust pipes closer to the engine. The valve angle was widened from 80 to 90 degrees (later 100 degrees); the valve diameter was also changed. It is worth noting that the exhaust valves, which were sodium cooled, were 15 per cent smaller in diameter than the inlet ones. All eight valves were supported by helical springs, enclosed by cylindrical tappets that were in direct contact with the camshafts.

Passoni also experimented with various crankshaft types: a forged one-piece item, necessitating the use of split big-ends and split edges for the big-end rollers; a built-up version in several pieces that were pressed together; and a built-up type assembled by the Hirth process. Some of these cranks were built by Gilera itself; whilst others were built in Germany.

Eventually it was determined that built-up cranks were best, having a life of

Grand Prix Glory

between 50 and 100 hours under racing conditions. A total of no less than six main bearings supported the crankshaft, and the whole engine assembly benefitted from the employment of both bearings and bushes of more than adequate dimensions for their particular task. The gearbox was given a fifth ratio, mainly in the interests of providing a lower first gear, thus restricting the use of the clutch.

Passoni also carried out experiments with battery/coil ignition, but ultimately it was found that the latest type of Lucas rotating-magnet magneto was superior. The factory claimed 64bhp at 10,500rpm.

Changes to the cycle parts were restricted to shortening the frame, narrowing the rear forks, fitting a streamlined cowling and the fitment of a more potent (and larger) twin leading-shoe front brake.

LIBERATI JOINS THE TEAM

In riding terms there had also been changes. Dale and Pagani had been signed

Reg Armstrong before the start of the 1954 Senior TT. Much of the actual race, shortened from six to four laps, was in heavy rain. The Irishman came home fourth at an average speed of 85.63mph (138km/h), losing a place as a result of a fourth lap pit stop.

Over 100,000 spectators crowded around the seven-kilometre Hedemora circuit to witness the Swedish Grand Prix on Saturday and Sunday 17/18 July 1954. The fastest lap was set by Reg Armstrong in the 500cc race at 102.52mph (165km/h), but he dropped back to finish third after refuelling on the twenty-seventh lap of the 130.48-mile (210km) race.

53

by rivals MV, leaving Duke and Armstrong, plus Libero Liberati, who mainly raced at home that year. In addition, Frenchman Pierre Monneret continued to receive limited assistance. However, his set-up proved enough, as the Arcore marque dominated the World 500cc Championship series yet again and in the process won all but three rounds.

The original nine rounds of the 1954 series were in fact reduced to eight that actually counted towards the Championship. This was as a result of the Ulster GP race distance being reduced below the required FIM limit due to inclement weather.

Monneret had given Gilera a head start by winning on home soil in the French GP at Reims. Afterwards Duke took over, winning five rounds on the trot. At season's end his victory margin over the runner-up, Norton's Ray Amm, was a clear twenty points.

At Monza, in September, Passoni introduced a new, streamlined shell (often referred to as the 'dustbin' fairing); it was a taste of things to come.

THE 1955 SEASON

For 1955 the four-cylinder Gilera was largely as it had been the previous year, except for a large cooling duct for the front brake hub, and the adoption of the streamlined fairing for all the team bikes. However, right from the start, some riders openly voiced their reservations about this latter device, claiming it made the machine more difficult to handle, particularly in high winds.

The seasonal 'roundabout' of riders didn't affect Gilera too much with the main riders only participating in their local events, such as Monneret (France), Martin (Belgium) and Veer (Holland).

THE DUTCH TT RIDERS STRIKE

But the year was to be remembered by Gilera fans for all the wrong reasons, and it was also to prove Geoff Duke's last full Championship season for the factory. Mid-season, with Duke and Armstrong dominating much of the proceedings, a strike by privateer riders at the Dutch TT ended with Duke and Armstrong both being barred from international events for the first half of 1956 – even though they had not taken part in the demonstration. Their only 'crime' being that they had argued the case for the privateers with the organization in a reasonable fashion!

There were eight rounds of the 500cc World Championship during 1955. Of these Duke not only took the fifth – his third and last for Gilera, but also won in France, the Isle of Man, Germany and Holland. Other Gilera victories were gained by Armstrong (Spain) and Colnago (Belgium).

The only real mechanical problem experienced that year centred on a spate of broken valve springs, later traced to faulty heat treatment.

BRITISH SHORT CIRCUITS

After the GP season was over, Duke raced a factory Gilera four at a quartet of British short circuits – at two of these he was beaten by a youngster riding a single cylinder Norton, John Surtees. This was noted by Count Domineco Agusta who promptly signed the promising Englishman for his MV team.

Grand Prix Glory

(Right) *Libero Liberati winning the first round of the Italian Senior Championships, 19 March 1955. This was the first year Gilera used the full 'dustbin' aluminium streamlined shell. Note the massive side vents to allow cooling air to reach the cylinders. The venue is the Circuit di Napoli.*

(Left) *Reg Armstrong rounds the Governor's Bridge dip during the 1955 Senior TT. Armstrong finished second to the race winner Geoff Duke (also riding a Gilera), averaging 96.76mph (156km/h).*

The Gilera fours of Alfredo Milani (40) and Giuseppe Colnago (38) battle for the lead at Senigallia, 31 July 1955. Milani won the race. Of the four Milani brothers, Alfredo was the most talented and stayed with Gilera throughout his racing career.

Grand Prix Glory

Geoff Duke – Champion of Champions

Geoff Duke was born in St Helens, Lancashire on 29 March 1923, the son of a baker. In 1942, at 19 years old, he volunteered for the army and, after a spell as a trainee mechanic, became a despatch rider with the Royal Corps of Signals. It was during his time in the army that he met men such as Hugh Viney (AJS trials star) and Freddie Frith (who later won the first 350cc racing World Championship on a Velocette).

Geoff Duke was demobbed in July 1947 and, on leaving the army and returning to St Helens, his first move was to purchase a new BSA B32 trials machine, his first victory on the machine coming soon afterwards. This success – and a good slice of luck – saw him meet up with Artie Bell, then the number one rider in Norton's road-racing squad. This was to result in the offer of a job at Norton (he was about to join AMC, the makers of AJS at Matchless) and the chance to ride the newly developed 500T trials model.

The Norton job brought about an entry of a standard 348cc Manx Norton model for the 1948 Junior Manx Grand Prix. But the young Duke was destined to retire half-way through the race with a split oil tank whilst leading his first ever road race!

Next came a third place in the 1949 Irish North West 200 on the same Manx, followed the next month by victory on a production International model in the Senior Clubmans TT. Then came more glory with a win in the Senior Manx Grand Prix – and runner-up in the Junior event following a crash. These successes came to the attention of Norton's race team supremo, Joe Craig, the result being that Geoff was promoted to the full factory works team for the 1950 season.

After making a winning début on the new McCandless-framed 500 Featherbed Manx at Blandford, a series of tyre problems meant that Norton didn't enjoy a very good 1950 season, but Geoff went on to score a highly impressive 350/500cc World Championship double the following year – and retained the 350cc title in 1952. But by now the four-cylinder Italian Gilera and MV models were fast outpacing the British single-cylinder Nortons. The result, having first considered a move into four wheels, was that Geoff jumped to sign for Gilera in the spring of 1953.

Geoff Duke then went on to the same level of achievement as he had done at Norton by winning a trio of World titles. But this time they were all in the blue-riband 500cc category for three consecutive seasons: 1953, 1954 and 1955.

Politics and accidents did much to restrict his efforts in 1956 and 1957, and then Gilera quit. This left Geoff on his own and he rode a mixture of Norton, BMW and Benelli machinery for the next two years before finally deciding to hang up his leathers after scoring a trio of wins at Locarno, Switzerland.

However, although he was to make his home in his beloved Isle of Man, Geoff Duke was not to quit the motorcycling scene; far from it, in fact.

One of his first moves was to visit Japan where he was given a hero's welcome as a guest of the Suzuki Motor Company. There were also some moves towards four wheels once again, but a crash in Sweden convinced Geoff that his future did not lay in this direction.

Later came the ill-fated return by Gilera through the Scuderia Duke team and, later still, an involvement with the British Royal Enfield Company, which saw Geoff helping with the design and development of the RE5, a 250 two-stroke, single-cylinder, production racer. Finally, in 1965 the Duke touch was applied to the organization of the ISDT (International Six Days Trial) held in the Isle of Man.

With the advent of the classic scene, which began at the end of the 1970s, Geoff was to find himself parading the Gilera four once again, much to the delight of enthusiasts worldwide.

Geoff Duke, the winner of three consecutive 500cc world championship titles with Gilera in 1953–55. Not just a champion, but a great ambassador for the sport.

Grand Prix Glory

(Top left) *Gilera team bikes in the Monza paddock, Italian Grand Prix, September 1953.*

(Top right) *Reg Armstrong winner of the 500cc German GP, Sunday 22 July 1956. Held over the superb Solitude circuit near Stuttgart, Armstrong won at an average speed of 92.02mph (148km/h) from Masetti's MV. This came after the sensational retirement of Geoff Duke, Bill Lomas (Moto Guzzi) and local hero, Walter Zeller (BMW).*

(Bottom left) *The 1955-type Gilera four. This superb drawing was commissioned by Gilera from the artist Cavara.*

The end of 1955 saw another significant change, with the long-serving Piero Taruffi giving up his role of team manager to concentrate on his four-wheel efforts. His place was taken by the next generation Gilera, the young and very enthusiastic Ferruccio, then 25-years-old.

THE FIM BAN

The FIM ban meant that Gilera was deprived not only of Duke and Armstrong, but also Colnago and Alfredo Milani. This meant that the company could not contest the first two rounds. As these were both won by Surtees and MV in a six-round series it was as good as over before Gilera even fired their engines. For the record book Surtees won another round in Belgium, before putting himself out of action after falling in Germany and fracturing his arm. Meanwhile Duke and Gilera had to be content with victory in the final round at Monza (where the Arcore fours annexed the first four places); Armstrong also won in Germany. However, in retrospect it had been, at least on past results, a bad year.

PASSONI WORKS ON

On the technical front Passoni had carried out more changes during the winter months. The frame was strengthened, a revised dustbin fairing with improved aerodynamics, an exhaust system which now included four megaphones (previously straight pipes were employed), and more power – up to 70bhp at 11,000rpm. An unfortunate by-product of these improvements was more weight, which had jumped to 330lb (150kg).

MECHANICAL GREMLINS STRIKE

Gilera also suffered a few mechanical problems to add to its woes, highlighted by Duke's retirement in the Belgium GP with a shattered piston (which was later traced to poor quality fuel – or at least that was Gilera's excuse).

Geoff Duke tested his Gilera four with experimental front and rear streamlining at Monza during practice for the 1956 Italian Grand Prix.

Four-Cylinder 350GP (1956–1963)

Engine:	Air-cooled, four-stroke, dohc, across-the-frame four-cylinder
Bore:	46mm
Stroke:	52.6mm
Displacement:	349.66cc
Compression ratio:	14:1
Maximum power (at crank):	49bhp @ 11,000rpm
Lubrication:	Wet sump
Ignition:	Magneto, Lucas
Fuel system:	4 × Dell'Orto SS 22 or 25mm
Primary drive:	Gear
Final drive:	Chain
Gearbox:	Five-speed
Frame:	Duplex, steel cradle
Front suspension:	Telescopic fork, fully enclosed
Rear suspension:	Twin shock, swinging arm
Front brake:	20mm drum 2LS
Rear brake:	200mm drum, 2LS
Wheels:	19in front and rear
Tyres:	Front 3.00 × 19; Rear 3.25 × 19
Wheelbase:	57in (1,450mm)
Dry weight:	320lb (145kg)
Fuel tank capacity:	*3.3imp. gal (15ltr)
Top speed:	149mph (240km/h)

*Some machines fitted with larger tanks for longer races

THE 350 FOUR ARRIVES

The only positive thing to come out of what had been a less-than-happy year, came at the final round at Monza, where not only did Gilera, as recorded earlier, take the first four places in the 500cc race, but the company also débuted a new 349.66cc (46 × 52.6mm) version of its four-cylinder design. With a choice of either 22 or 25mm Dell'Ortos, depending upon the circuit, the smaller multi produced 49bhp at 11,000rpm.

Two of the smaller Gileras appeared on the start line and, although Duke retired, Liberati kept going for the whole race distance to win comfortably from Dickie Dale's Guzzi single. Liberati's fastest lap of 1m 52s, 114.84mph (185km/h), compared with the 500cc lap record (set jointly by Duke and Liberati) at 1m 50.4s, 116.51mph (187km/h).

Just to seal things for Gilera, they made a victorious return to the sidecar class, with the brothers Milani winning as they pleased from a field of BMWs and Nortons. The same pairing returned a year later to repeat the act but otherwise, as recounted earlier, Gilera had retired from the three-wheel action in 1952.

AND A 125 TWIN

Not only had Passoni been working on the creation of a smaller four, but also a brand

Grand Prix Glory

Geoff Duke displays all the style that made him the most successful rider of his era, during his victorious ride in the 500cc Italian GP, 9 September 1956. There were an amazing six Gilera fours in the race – Duke, plus Reg Armstrong, Pierre Monneret, Alfredo Milani, Libero Liberati and Giuseppe Colnago.

new 125 twin, which had been designed during 1955. This was first seen at Monza in May 1956, with lightweight jockey Romolo Ferri aboard. Its start-to-finish victory over a mass of FB Mondials and MV Agustas made the opposition take serious notice, but the Grand Prix career of this jewel-like 124.656cc (40 × 49.6mm) dohc machine was destined to be blighted by mechanical problems. In fact, it only ever won a single GP, the German at Solitude in July 1956.

ARMSTRONG RETIRES

The year was brought to a close by two off-circuit incidents that were to have a significant effect on Gilera's racing future, but in vastly different ways. In September, after racing at the Avus circuit in Berlin, Reg Armstrong announced his retirement; in the following month, Ferruccio Gilera, then 26-years-old, suffered a fatal heart attack while visiting the company's Argentinian subsidiary in Buenos Aires.

Twin-Cylinder 125 GP (1956–1957)

Engine:	Air-cooled, four-stroke, dohc, parallel twin, with cylinders inclined 30 degrees from vertical
Bore:	40mm
Stroke:	49.6mm
Displacement:	124.456cc
Compression ratio:	11:1
Maximum power (at crank):	18bhp @ 12,000rpm
Lubrication:	Wet sump
Ignition:	Battery coil
Fuel system:	2 × Amal GP or Dell'Orto SS 22mm carbs
Primary drive:	Gear
Final drive:	Chain
Gearbox:	Six-speed
Frame:	Duplex, steel cradle
Front suspension:	Telescopic fork, exposed stanchion
Rear suspension:	Twin shock, swinging arm
Front brake:	Full-width drum, 2LS
Rear brake:	Full-width drum, 2LS
Wheels:	Front and rear 18in
Tyres:	Front 2.00 × 18; Rear 2.50 × 18
Wheelbase:	48in (1,230mm)
Dry weight:	209lb (95kg)
Fuel tank capacity:	3.3imp. gal (15ltr)
Top speed:	118mph (190km/h)

Note:
Originally raced with dustbin streamlining, 1956 and 1957. Then retired. Later raced in Italian Championship series in 1966–1967 with dolphin fairing.

Lightweight star Romolo Ferri at Monza in September 1956 on the fully streamlined 124.65cc (40 × 49.6mm) dohc 125GP twin.

Grand Prix Glory

Libero Liberati giving the new 350 four a convincing victory début at Monza on the 9 September 1956. He also set a new class lap at 114.84mph (184.8km/h).

(Below) Geoff Duke and spanner man Giovanni Fumagelli at Monza in September 1956. Fumagelli was probably the best known of all the Arcore company's race mechanics.

MCINTYRE JOINS THE SQUAD

Armstrong's retirement was to lead to the Scot, Bob McIntyre, joining the squad, while Ferruccio's unexpected death robbed the team of a most enthusiastic leader (it was he who had inspired Passoni to design the new 125 twin and 350 four-cylinder models). As if the tragedy was not enough, it was also instrumental in draining Giuseppe Gilera of the determination and drive which had created what was, at the time, one of the truly great racing teams in existence. It was almost as if he had lost much of the will to live, so great was the loss of his only son. As someone who has also lost a son, I truly know just what Giuseppe Gilera went through: it is heartbreaking.

Giuseppe Gilera

Born in a small village near Milan on 21 December 1887, Giuseppe Gilera, and his younger brother Luigi, did more than most to transform the face of motorcycle racing. For it was Gilera, rather than any other factory, who introduced the across-the-frame, four-cylinder to Grand Prix racing. Others such as MV Agusta, Benelli and Honda took up the layout, but it was Gilera who got in first.

Others may have noticed the possibilities of the machine raced by the Rome-based Rondine squad during the mid-1930s, but Giuseppe Gilera was the one who snatched the opportunity by actually purchasing the design rights.

The Rondine became the Gilera in 1936. And after almost three years of development became the Champion of Europe in 1939. However, even then the reliability had to wait until the post-war years. Even so, on the very eve of war, as long as the 500 water-cooled supercharged Gilera kept going, nothing on two wheels could touch it.

Gilera was also a talent scout and gifted engineer. For starters he not only coped when Ing. Remor left for MV Agusta at the end of 1949, but also helped to develop Remor's air-cooled, normally aspirated four into World Championship material. In this he was greatly helped by the procurement of the services of Geoff Duke, whom he finally signed early in 1953 (after making his original advances as early as 1950).

Duke was later to recall: 'He was a fabulous man – a real gentleman, and most generous, not just a boardroom figurehead. You couldn't fault him on a single thing.'

Duke quit Norton to sign for Gilera. And in his first year, 1953, won the 500cc World title – the first of his three consecutive individual World crowns for the Arcore factory. But, sidelined by injury, he then missed Gilera's crowning glory – the first official 100mph (160km/h) lap of the legendary 37.75 mile (60.75km) Isle of Man circuit and the Junior-Senior double, notched up by his nominee, Bob McIntyre, in the 1957 Golden Jubilee Tourist Trophy.

Then, after Gilera announced its retirement (along with Moto Guzzi and FB Mondial), McIntyre went to Monza in November 1957 and broke the One-Hour World Record on a 350 four – a 500 couldn't be used because of the poor condition of Monza's banking!

But even though Gilera had pulled out, its influence was felt in the years that followed, first by MV Agusta and then by the Japanese Honda team.

But, as for Giuseppe Gilera himself, his later years were to be filled with great disappointment. First, his beloved and only son, Ferruccio (then only 26 years old) died of a heart attack whilst on a business trip to the Argentinean Gilera subsidiary in October 1956. During the 1960s he was forced to fight an uphill battle in a declining motorcycle market, to keep the company he had created afloat. It proved to be a feat that, ultimately, even he was unable to achieve, with Gilera being taken over in 1969 by the Piaggio organization.

Only a few short months later, on 21 November 1971, Giuseppe Gilera passed away. He was 83 years old.

Father and son: Guiseppe Gilera and his only son Ferruccio at the Belgian GP in July 1956. The latter's sudden death only a few days later whilst on a business trip to the Argentine left Guiseppe a broken man.

Four-Cylinder 500GP (1956–1966)	
Engine:	Air-cooled, four-stroke, dohc, across-the-frame four-cylinder
Bore:	52mm
Stroke:	58.8mm
Displacement:	499.49cc
Compression ratio:	13:1
Maximum power (at crank):	70bhp @ 10,500rpm
Lubrication:	Wet sump
Ignition:	Magneto, Lucas
Fuel system:	4 × Dell'Orto SS 25 0r 28mm
Primary drive:	Gear
Final drive:	Chain
Gearbox:	Five-speed (1966 seven-speed)
Frame:	Duplex, steel cradle
Front suspension:	Telescopic fork, fully enclosed
Rear suspension:	Twin shock, swinging arm
Front brake:	20mm drum, 2LS
Rear brake:	200mm drum, 2LS
Wheels:	19in front and rear
Tyres:	Front 3.00 × 19; Rear 3.50 × 19
Wheelbase:	57in (1,450mm)
Dry weight:	330.5lb (150kg)
Fuel tank capacity:	*3.3imp. gal (15ltr)
Top speed:	161.5mph (260km/h)

* Some machines fitted with larger tanks for longer races

1957 – A SEASON OF HIGH DRAMA

And so to 1957, destined to be a year of high drama. However, it didn't get off to a good start for Gilera, with Duke putting himself out of action after casting his machine away in spectacular fashion during the season's opening meeting, the International Shell Gold Cup at Imola. At the same meeting, new boy McIntyre's machine was forced out after electrical gremlins set in.

Then came the World Championships. The first round was at Hockenheim and was marked by wet roads and records being shattered in every race. Although McIntyre set new records for both the 350 and 500cc races, it was his team-mate Liberati who actually won both events. In the 350cc race McIntyre fell whilst leading; then in the 500cc class he first experienced his machine running on only three of its four cylinders and, after stopping at his pit, charged back through the field to find himself a mere *20 yards* astern of race winner Liberati at the chequered flag.

TT DOUBLE

The 'Flying Scot' then roared back to score one of the most famous double victories ever, when he annexed both the Junior and Senior Golden Jubilee TTs – the latter staged over no less than eight laps, 302

Grand Prix Glory

(Top left) *Gilera's renowned mechanic Giovanni Fumagelli, seen here at the TT in June 1957.*

(Above right) *The Australian Bob Brown was drafted into the Gilera team for the Golden Jubilee TT of June 1957. A former Sydney taxi driver, Brown came to Europe in 1955. Right from the start he put in some highly impressive performances on Matchless, AJS and NSU machinery. A friend of Geoff Duke's, it was the latter's Imola accident that led to the TT ride.*

(Bottom left) *A pensive Bob McIntyre pushes his smaller Gilera four before the start of the 1957 Junior TT. He went on to score a magnificent double TT victory (350 and 500cc) in the Golden Jubilee of the event.*

miles (486km), of the famous 37.75 mile (60.75km) mountain circuit. He also became the first man in history to lap the circuit in excess of 100mph (160km/h). As a point of interest, the Australian Bob Brown was drafted into the squad for the Isle of Man, rewarding Gilera with a pair of fine third places.

MCINTYRE IS INJURED

Next came the Dutch TT, where McIntyre suffered an accident, the consequence being a neck injury that was to plague him for the remainder of the year. And even though both he and Duke were fit enough to ride in the next two rounds (Ulster and Italy), Liberati was able to collect enough points to become Gilera's leading points scorer in both 350 and 500cc categories, resulting in the Italian's becoming runner up in the 350cc and champion in the 500cc.

Virtually no changes had been carried out from the 1956 models, this being a sign

Gilera's Last Champion

Even after his 1957 500cc World Championship title, to many people outside Italy, Libero Liberati remained a mystery man.

When he took the title at the final round at Monza he was almost 32 years old and had been racing for twelve years. In fact he had enjoyed the backing of the Gilera factory since 1950 – and first rode a four the following year, so in reality his rise came slower than most.

Besides his racing contract with Gilera he was also one of their dealers, with a showroom in Terni, near Rome.

The son of a butcher, he left school at the age of fourteen, and began work as a mechanic. Even at this young age he was keen on motorcycling and had already had his first ride – on a small capacity Mas four-stroke single – some two or three years earlier.

Then came the war, but when competition resumed after the conflict, Liberati, then twenty, made his début on a Moto Guzzi flat single in a local hill climb. That model soon gave way to a Guzzi Condor and, before 1946 was out, came the first of his victories.

Next a Gilera Saturno Sanremo was obtained. At first Liberati raced privately, but after a string of excellent results, he eventually received help from the Arcone factory, followed by a factory Saturno for the 1950 season.

At Monza in September 1951 came his début on a four at the *Grand Prix des Nations*.

During 1952 Gilera entered Liberati in the Italian and selected World 500cc Championship

Libero Liberati winning the 1955 500cc Senior Italian Championship on his works four-cylinder Gilera. He was also Gilera's last racing championship title holder after becoming the 500cc world champion in 1957.

Gilera's Last Champion

rounds. He rode abroad in Spain, Holland and Switzerland. It was at the latter circuit in Berne that he was to crash and sustain a severe injury which left him with a 'bent arm' riding style for the remainder of his career.

One of the reasons why he was so little known outside his homeland was that he never raced in the Isle of Man (then the world's premier event) and only once in Ulster.

After winning the 1955 and 1956 Italian 500cc Senior Championship titles, Liberati scored his first Grand Prix victory on the new 350 four-cylinder model in September 1956; he was also runner-up behind Geoff Duke in the 500cc event.

In his Championship year, Liberati took victories at Hockenheim, Dundrod and Monza – he also won at Spa Franchomps, but was excluded from the results for entering the race on team-mate Bob Brown's machine. A 350cc victory was notched up at Hockenheim on the smaller four.

Some critics said that if Duke and McIntyre hadn't suffered injuries, or that if so-and-so hadn't retired, Liberati would not have taken the title, but in racing, 'points win prizes' and that is exactly what Libero Liberati did – he scored the most points.

After Gilera withdrew from racing at the end of 1957, Liberati continued to compete in minor events on his own Gilera Saturno Corsa single. It was a measure of his sportsmanship and enthusiasm for the marque that he persevered on what was, by then, an outdated machine.

Sadly, it was to be on his faithful Saturno that Libero Liberati was to meet his death on Monday 5 March 1962, whilst testing the Gilera on public roads near his home town of Terni in readiness for the coming season's racing.

---

that all was not well within Gilera itself. As Geoff Duke was to recall later, the vital spark seemed to be missing during 1957, although to outsiders this wasn't noticeable at the time.

GILERA ANNOUNCES ITS RETIREMENT

When the chance came to withdraw, after almost fifty years of competitive motorcycle sport, Giuseppe Gilera grabbed it with both hands. This took the form of a tripartite agreement between Guzzi, Mondial and Gilera, all of whom announced their withdrawal from Grand Prix racing at the end of the year. For Guzzi and Mondial, the matter was purely financial, but for Giuseppe Gilera it was not only a matter of cost, but also a consequence of the air of depression hanging over the company caused by the death of his son.

RUMOUR AND COUNTER RUMOUR

Before finally mothballing the bikes, Gilera went on a record-breaking spree at Monza, (*see* Chapter 4), including McIntyre's historic hour of triumph. There followed a period of almost five years when rumours abounded about the marque's possible comeback. The real truth was that no such comeback had been seriously considered, even though a whole string of riders and sponsors had besieged the factory with requests to loan them machines.

OULTON PARK, AUGUST 1962

Then, in August 1962, the former Gilera star, Bob McIntyre, crashed his 500cc Manx Norton at Oulton Park and was fatally injured. A remembrance meeting

The 175 Twin (1957)

Even though, unlike the fours and the 125 twin, the 175 twin wasn't a Grand Prix bike, it was nonetheless an interesting effort.

Born out of the need for publicity to prop up flagging domestic sales, the 175 engine was based around the two inner cylinders of the factory 350 four Grand Prix design, and utilized the same 46 × 52.6mm bore and stroke dimensions.

Designer Ing. Franco Passoni completed his brief over the winter of 1956–1957. A total of ten machines were constructed, their purpose being primarily to take part in the long-distance road events such as the Moto Giro (Tour of Italy) and Milano–Taranto classics. Whereas Gilera had already won the latter event outright in both 1955 and 1956 – Bruno Francisci (four-cylinder GP) and Pietro Carissoni (Saturno Piuma) – the Arcore marque had not had much success in the lightweight categories. It was in the smaller classes that rivals such as Ducati, Laverda and Morini had been victorious – and benefitted from increased sales of their standard production models.

The cylinders of the 175 were not so steeply inclined at 20 degrees, as those on the smaller capacity, Passoni-designed 125 twin-cylinder Grand Prix engine.

Another feature of the 175 compared to the 125 was the much longer inlet spacers between the heads and the 24mm Dell'Orto SSI carburettors.

One could have expected Ing. Passoni to simply utilize the chassis from the Grand Prix machine, but he didn't. There was a new frame which differed from the smaller mount by featuring a single top tube and no longer used two of the GP machine's bracing tubes. However, the forks, rear suspension and brakes were 'borrowed' from the 125.

Clearly based on the earlier 125GP twin, the 174.8cc (46 × 52.6mm) dohc 175 F2 of 1957 was intended for long-distance road events, such as the Milano–Taranto and Giro d'Italia (Tour of Italy).

Grand Prix Glory

The 175 Twin (1957)

Producing 23bhp at 11,200rpm the double overhead cam engine was of full unit construction design with a five-speed gearbox and multi-plate clutch. On paper this looked a competitive enough package to ensure success. However, the machine's 105mph (170km/h) was some 7–9mph (12–15km/h) down from its 125 brother – its lack of streamlining at least being a contributing factor.

The 1957 Moto Giro (the last such event due to the public and press outcry following the Mille Miglia car fatalities) saw Gilera enter several riders on the new 175 twin, but the best it could do was a fifth in its class with the experienced Pietro Carissoni aboard. A win by Remo Venturi with a works-entered MV Agusta didn't make things easier – as MV were Gilera's main rivals in the Grand Prix arena.

Later that year, with the axing of the planned Milano–Taranto, the machines were entered in several rounds of the Italian Formula Two Championship series. Again, results were less than glittering; the machine proving reliable but simply not quick enough. The only track success was in the Italian Junior Championship, with Giancarlo Muscio claiming third spot in the 1957 title race.

Finally, in November 1957, Gilera took the 175 twin, together with its other racers, all suitably clad in full enclosure, for a record breaking spree at the nearby Monza Autodrome. Here, at last, the 175 performed as Passoni and Giuseppe Gilera had always hoped it would, setting a number of new speed records for various distances.

Rider Romolo Ferri established several important new benchmarks in both the 175 and 250cc categories. These included:

Standing Kilometre – 27.7 seconds – 81.97mph (132km/h)
100 kilometres – 129.39mph (208km/h)
One-Hour – 129.57mph (208.5km/h).

The record breaking exploits were to signal the end of Gilera's participation in both racing and speed record breaking – at least until the early 1960s.

---

was staged in mid-October at the same circuit and, as a mark of respect, the Arcore factory sent one of its 1957 'dustbin-faired' fours over for Geoff Duke to parade round the circuit. It was this event which was to trigger a Gilera comeback.

THE 1963 COMEBACK

In early March 1963 came the news that the famous Arcore fours would race again, under the 'private' Scuderia Duke banner with Derek Minter and John Hartle as riders. After testing at Monza the team returned to Britain. Their first appearance in this country came at Silverstone on 6 April.

I was there that day and can still remember the expectant buzz all around the Northamptonshire circuit, which everyone in the vast crowd shared. But although Minter duly won the race, it was the sight of the Norton-mounted Phil Read splitting the Italian multis on what was, after all, a speed circuit (although Hartle finally got the better of the Norton privateer), which put the question on everyone's lips – would Gilera do as well as the pundits had previously expected?

SET-BACK FOR MINTER

And so it transpired; Minter was badly injured before the GPs even started, after

Twin-Cylinder 175 Formula Two (1957)

Engine:	Air-cooled, four-stroke, dohc, parallel twin, with cylinders inclined 20 degrees from vertical
Bore:	46mm
Stroke:	52.6mm
Displacement:	174.8cc
Compression ratio:	10.7:1
Maximum power (at crank):	23bhp @ 11,200rpm
Lubrication:	Wet sump
Ignition:	Battery coil
Fuel system:	2 × Dell'Orto SSI 24mm carbs
Primary drive:	Gear
Final drive:	Chain
Gearbox:	Five-speed
Frame:	Duplex, steel cradle
Front suspension:	Telescopic fork, exposed stanchion
Rear suspension:	Twin shock, swinging arm
Front brake:	Full-width drum, 2LS
Rear brake:	Full-width drum, 2LS
Wheels:	Front and rear 18in
Tyres:	Front 2.00 × 18; Rear 2.75 × 18
Wheelbase:	52in (1,320mm)
Dry weight:	253.5lb (115kg)
Fuel tank capacity:	4imp. gal (18ltr)
Top speed:	105mph (170km/h)

Cover of a sales brochure produced during the late 1950's proclaiming Gilera's ten World Championship titles. The rider is Bob McIntyre, the year 1957.

Libero Liberati (left, in coat) with Bob McIntyre at Monza in November 1957. As a strange quirk of fate, both were to meet with fatal accidents in 1962.

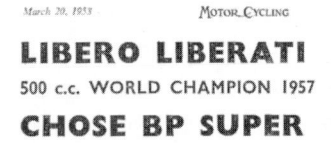

Grand Prix Glory

(Top left) *This full page BP advertisement appeared on the 20 March 1958 issue of* Motor Cycling, *giving details of Liberati's 1957 500cc World Championship success.*

(Top right) *The Gilera race shop gathering dust in the autumn of 1962. After five long years away from the circuit the lack of use is evident from the disorganized remains of what were once the most glamorous machines on the planet. A few short months later this mass of mechanical debris was to be readied again for battle.*

(Bottom left) *Ing. Franco Passoni showing racer/journalist Roberto Patrignani one of the 174.8cc (46 ×52.6mm) Formula Two racing engines during the winter of 1962/1963 as thoughts returned to a Gilera comeback.*

Grand Prix Glory

(Left) *Geoff Duke (left), Derek Minter (centre) and journalist Charlie Rous, together at Monza during testing which preceded Gilera's comeback under the Scuderia Duke banner in 1963. Unfortunately, an accident at Brands Hatch (whilst racing his own Norton) sidelined Minter for the majority of that year's racing.*

an accident on one of his own Norton singles at Brands Hatch in early May, which resulted in Read being brought into the team as a replacement.

The next problem arose at the first Grand Prix, the West German at Hockenheim. Here the smaller four was totally outclassed for speed, not just by the full-factory-backed Honda fours and Bianchi twins, but also by the pre-production Honda CR77 twin. Even though Hartle subsequently brought a 350 Gilera home second in the Junior TT, the writing was clearly on the wall and the smaller four was withdrawn so the team could concentrate on the 500cc class.

(Below) *Monza Autodrome, March 1963. The Gilera racing effort is reborn through Scuderia Duke. A group of journalists from around the world watch as John Hartle prepares to start one of the fours.*

Gilera mechanics readying the fours for testing at Monza in March 1963.

73

Grand Prix Glory

John Hartle (Gilera) and Phil Read (Norton) scrap during the Hutchinson 100 at Silverstone in early April 1963. Although Hartle won the argument and made it a Gilera one-two (Minter won), it was clear to see that Gilera wasn't going to have things all its own way.

A week later at Oulton Park Derek Minter thrilled a mammoth crowd of over 50,000 by clipping nearly two seconds off the existing lap record with a fabulous 91.86mph (147.8km/h).

Geoff Duke, with the smaller Gilera four, at the 1963 Isle of Man TT, with chief mechanic Giovanni Fumagelli. Rider John Hartle finished runner-up to race winner Jim Redman's Honda four. A good performance, but not enough to satisfy the expectations of those who imagined the Arcore factory would resume the domination of earlier days.

Grand Prix Glory

DUTCH TT VICTORY

However, hopes were not matched by results, and, except at an early non-Championship event at Imola, the bigger Gileras simply could not match the lone MV Agusta of Mike Hailwood. Hartle gave the Scuderia Duke equipé its only GP victory in Holland, but this was due to Hailwood's MV blowing up on the second lap of the race. Minter had returned to the squad by August, but this did not seem to signify any real improvement in its fortunes.

UNCERTAINTY AGAIN

There then followed a period of uncertainty as to whether or not the company would race in 1964. It carried out yet more testing at Monza – this time with several Italian riders, including Gilberto Milani, Franco Mancini and Renzo Rossi. John Hartle was also on hand, riding one of his fours equipped with a British Reynolds leading-link fork assembly. Still no official statement was issued on future plans.

(Top left) *Tank-off view of the four as it was in 1963.*

(Top right) *Mallory Park 'Race of the Year' September 1963. The Gilera fours of Derek Minter (4) and Phil Read (6). The race was won by Mike Hailwood (MV), followed by Minter. Read was fourth.*

(Bottom left) *In October 1963, Gilera invited various Italian riders to test the fours at Monza. Included (left to right) were Gilberto Milani (Aermacchi) Franco Mancini (Motobi) and Renzo Rossi (Bianchi).*

BENEDICTO CALDARELLA

The first round of the 1964 World Championship series was staged at Daytona, Florida, the 500cc event taking place in early February. This created something of a sensation when, for no less than 75 miles (120km), Mike Hailwood, then the undisputed master of the sport, fought a wheel-to-wheel battle with the little-known Argentinian Benedicto Caldarella.

The champion was on his MV, the challenger on a Gilera. Only gearbox trouble finally removed Caldarella from this amazing battle. For lap after lap of the 3-mile (4.8km) American circuit, the two Italian fours duelled for the lead in the closest 500cc battle for years. It later transpired that the Gilera was one of the Scuderia Duke bikes which had been sent to the Argentine at the end of the previous year.

As if to prove the United States GP was not a one-off, Caldarella went on to win the 500cc race at the international Imola meeting in April at record speed (Mike Hailwood did not take part).

Benedicto Caldarella, the one man in 1964 to challenge World Champion Mike Hailwood and his all-conquering MV Agusta.

Caldarella (46) and Remo Venturi (Bianchi twin) during their race-long dual at the Imola Gold Cup meeting in April 1964. The diminutive Argentine Gilera rider came out on top.

Benedicto Caldarella – South American Challenger

The way Benedicto Caldarella challenged the then undisputed master, Mike Hailwood, in their first ever encounter, during the 1964 500cc United States Grand Prix at Daytona, was enough to guarantee him a special place in Gilera folklore.

His Daytona performance was such, that for some 75 miles (120km) he fought tooth and nail with the MV star on one of the 1957-type Gilera fours. And, even though he was ultimately to retire from the race with gearbox gremlins, it was enough to earn the stocky Argentinian an invitation to ride for Gilera in the ensuing big-time European meetings. As recorded elsewhere, he didn't make much of a mark in the remainder of the World Championship series, but much of this was due to outside factors, rather than his own ability. However, in his comparatively few GP rides, Caldarella displayed sufficient brilliance to justify the label of 'South America's first motorcycle racing star'.

Benedicto Caldarella was a second-generation racer – like Hailwood, Surtees and many others. His father Salvador, one of a family of Sicilian emigrants who arrived in Argentina during the early part of the twentieth century, was a medical photographer by trade until his enthusiasm for racing and tuning motorcycles persuaded him to change his job during the 1940s. He later went on to establish a motorcycle dealership in Vincente Lopez.

In 1955, when Salvador was the 500cc champion of Argentina, eldest son Benedicto made his racing début at the tender age of fifteen, on a borrowed 150cc Gilera single. But enforcement of the eighteen-year, minimum age limit then in force put a stop to any repetition of that particular antic and he was forced to wait another three years until 1958 – the season in which Salvador decided to hang up his leathers – before going racing again, this time on a 500 Gilera Saturno.

Runner-up in the 500cc Argentinian Championship the following season, Benedicto won it in 1960. Not only this, but he also became South American Senior Champion, after a series of victories in neighbouring countries – one in a 100 mile (160km), 108-lap event staged over probably the most serpentine short circuit in the world, at Valparaiso, Chile.

Switching from the Saturno to a brand new Matchless G50 in 1961, he finished third in the National Championship. He went on to win the 1962 Argentine Grand Prix (counting towards the World Championship) on the very same bike. After another excellent season in 1963, including a third behind Hailwood and Jorge Kissling in the 1963 Argentine GP, he was provided with the four-cylinder Gilera with which he took on Hailwood at Daytona.

Benedicto Caldarella, seen here astride the four-cylinder Gilera he rode during the Centennial TT, Assen, in May 1998

THE CAMATHIAS SIDECAR

The factory also loaned the Swiss sidecar ace Florian Camathias an engine for record-breaking and 'selected' meetings at the beginning of 1964. After Camathias took victory in Spain, observers believed Gilera would mount a serious challenge for three-wheel honours and, together with Caldarella's performances, things looked bright. But a series of crippling strikes, together with other industrial problems, ensured that these hopes were soon dashed. The result was that the only other success came in the final round at Monza, where the diminutive Argentinian finished runner-up to Hailwood. Caldarella was some ten seconds adrift at the flag but did have the satisfaction of setting the fastest lap, at 121.08mph (195km/h).

STRIKES AND UNREST

The strikes and industrial unrest had only served to hasten the financial decline of the once-great marque and meant that any hope of Gilera competing had to be shelved. However, the Italian race organizers made substantial offers of financial assistance to Gilera as an encouragement to reappear in 1966 – the tactic had the desired effect and induced the factory out of retirement.

Gilera Sidecar Grand Prix Victories

1949	Ercole Frigerio/E Ricotti	Italian GP
1951	Ercole Frigerio/E Ricotti	Swiss GP
	Albino Milani/G Pizzocri	Italian GP
1952	Albino Milani/G Pizzocri	Swiss GP
	Ernesto Merli/D Magri	Italian GP
1956	Albino Milani/R Milani	Italian GP
1957	Albino Milani/R Milani	Italian GP
1964	Florian Camathias/R Foell	Spanish GP

Note:
All these victories were gained using the four-cylinder engine

Benedicto Caldarella (left) and Remo Venturi acknowledge the cheers after finishing first and second in the 500cc race at the international Imola meeting, April 1964.

Giuseppe Gilera (second from left, wearing hat) with a smiling Benedicto Caldarella in 1964.

The Swiss Sidecar ace Florian Camathias campaigned a four-cylinder Gilera-engined sidecar during the 1964 season – he also went record-breaking with the same outfit. But as with Cadarella's solo effort, industrial unrest at the Arcore factory put a spanner in the works.

Ing. Lino Tonti (formerly of Aermacchi and Bianchi, to name but two of his former employers) was hired in an attempt to update the now quickly aging machinery. Tonti proposed a selected number of modifications, the most noticeable being a new, slimmer faring and, more importantly from a technical viewpoint, a six-speed (and later, seven-speed) gearbox. In addition, the duplex front brake, first tested in 1957, was re-introduced.

However, all this activity was really of no avail as Gilera itself was heading still further down the road to financial disaster and thus was totally unable to fund any real challenge. The riders, Derek Minter and Remo Venturi, tried their best but scenes off the track transpired to make this, the final Gilera return, nothing more than a damp squib. A series of mechanical problems and a crash by Minter during TT practice only heightened the problems.

FINAL DAYS

The final appearance of the 500cc four-cylinder Gileras outside Italy came on an October's day at Brands Hatch, where Venturi was joined by ex-Suzuki star Frank Perris; however, the pair could only finish in mid-field. The final domestic turn-out came a couple of weeks later at Vallelunga, just north of Rome, where Venturi finished a lonely runner-up to Renzo Pasolini's Benelli, after race leader Agostini's MV Agusta retired.

Grand Prix Glory

Gilera's last British outing came at Brands Hatch on 9 October 1966, when former Suzuki star Frank Perris (8) and the Italian veteran Remo Venturi (4) rode a pair of 500 fours. Both could only finish mid-field – a sad end for a once great motorcycle.

Although both Perris and Minter tried to borrow bikes for the 1967 season, Giuseppe Gilera declined the advances. He had realized that one of the greatest racing motorcycles of all time had finally come to the end of the road. A few short months on, at the end of 1968, the Gilera company was forced to appoint a receiver. The following year the giant Piaggio organization took over and, although the Arcore marque was to survive, racing was not on the agenda.

The 1956 Italian Sidecar GP victor, Albino Milani. An outstanding speed superiority over the BMW and Norton opposition made for an easy victory. The Gilera star's winning average speed was 98.13mph (157.9km/h).

Gilera Solo Grand Prix Victories

Year	Rider	Class	Event
1949	Nello Pagani	500cc	Dutch TT
	Nello Pagani	500cc	Italian GP
1950	Umberto Masetti	500cc	Belgian GP
	Umberto Masetti	500cc	Dutch TT
1951	Umberto Masetti	500cc	Spanish GP
	Alfredo Milani	500cc	French GP
	Alfredo Milani	500cc	Italian GP
1952	Umberto Masetti	500cc	Dutch TT
	Umberto Masetti	500cc	Belgian GP
	Crommie McCandless	500cc	Ulster GP
1953	Geoff Duke	500cc	Ulster GP
	Alfredo Milani	500cc	Belgian GP
	Geoff Duke	500cc	French GP
	Geoff Duke	500cc	Swiss GP
	Geoff Duke	500cc	Italian GP
1954	Pierre Monneret	500cc	French GP
	Geoff Duke	500cc	Belgian GP
	Geoff Duke	500cc	Dutch TT
	Geoff Duke	500cc	German GP
	Geoff Duke	500cc	Swiss GP
	Geoff Duke	500cc	Italian GP
1955	Reg Armstrong	500cc	Spanish GP
	Geoff Duke	500cc	French GP
	Geoff Duke	500cc	Isle of Man TT
	Geoff Duke	500cc	German GP
	Geoff Duke	500cc	Belgian GP
	Geoff Duke	500cc	Dutch TT
1956	Romolo Ferri	125cc	German GP
	Reg Armstrong	500cc	German GP
	Geoff Duke	500cc	Italian GP
	Libero Liberati	350cc	Italian GP
1957	Libero Liberati	500cc	German GP
	Libero Liberati	350cc	German GP
	Bob McIntyre	500cc	Isle of Man TT
	Bob McIntyre	350cc	Isle of Man TT
	Libero Liberati	500cc	Ulster GP
	Libero Liberati	500cc	Italian GP
	Bob McIntyre	350cc	Italian GP
1963	John Hartle	500cc	Dutch TT

Riders to Have Raced Gilera Works Machines 1936–1967

Giordano Aldrighetti
Fernando Aranda
Reg Armstrong
Arciso Artesiani
Carlo Bandirola
Bob Brown
Benedicto Calderella
Florian Camathias
Giuseppe Colnago
Dickie Dale
Geoff Duke
Romolo Ferri
Tito Forconi
Bruno Francisci
Ercole Frigerio
Carlo Fumagalli
Claudio Galliani
Martino Giani
Luigi Gilera
Silvio Grassetti
John Hartle
Fritz Klaeger
Jacob Keller
Francesco Lama
Libero Liberati
Giovanni Lombardi

Felice Macchi
Leon Martin
Umberto Masetti
Massimo Masserini
Tullio Masserini
Cromie McCandless
Bob McIntyre
Ernesto Merlo
Albino Milani
Alfredo Milani
Georges Monneret
Pierre Monneret
Nello Pagani
Frank Perris
H. Adolfo Pochettino
Phil Read
Amilcare Rossetti
Dorino Serafini
Piero Taruffi
Silvio Vailati
Orlando Valdinoci
Marc Veer
Remo Venturi
Enzo Vezzalini
Ettore Villa

Machines
350/500cc fours
125cc twin
Sidecar four

Note:
Saturno single and 175 twin not included

4 Record Breaking

Two men are forever associated with Gilera's glorious achievements in the field of record breaking: Piero Taruffi and Bob McIntyre; both becoming the holders of the coveted One-Hour title pre-war and post-war, respectively.

PIERO TARUFFI

The Rondine was the forerunner of the Gilera four-cylinder and it was on a partially-streamlined Rondine that Taruffi began his record-breaking career in 1935. During his first session he raised the existing 500cc World Records for the Flying Mile and Kilometre to over 151.51mph (244km/h).

Held on the Firenze-Mare *autostrada*, designer Gianini and rider Taruffi had realized that drag was a crucial factor and equipped the machine with a fairing which, although the front wheel was uncovered, was pretty comprehensive nonetheless. Extending beneath the engine a long tail was added, enclosing the rear section of the machine as well as the majority of the front.

COUNT BONMARTINI

Count Bonmartini, owner of CNA and thus the Rondine at the time, had crashed whilst unofficially testing the bike a couple of days before Taruffi's successful bid. Although unhurt in the incident, the 'get-off' had been at well over 100mph (160km/h) and damage to the machine necessitated not only repairs to the streamlining, but also a new frame.

Shortly after this, Count Bonmartini sold out to the Caproni aeronautical empire and thus started the chain of events that, ultimately, saw Gilera take on the Rondine project.

Although post-war the Caproni organization was to take an interest in the motorcycle field, in 1936 its priorities were purely of an aviation nature; Taruffi was therefore instructed to find a buyer for the unwanted racers and record breakers. As Taruffi soon found out, this was to be a relatively easy task – and Giuseppe Gilera was the lucky buyer. Lucky, because here was a way to 'fast track' his racing ambitions. Taruffi was also asked by Gilera to join the company to work (and ride!) the ex-CNA four-cylinder machines.

GILERA INVOLVEMENT

Gilera's involvement also ensured that, at last, a serious development programme took place, from which was to rise one of the truly great record breaking motorcycle teams of all time.

Before leaving Caproni, Taruffi had been able to make use of the company's aerodynamic facilities, where he designed an all-embracing aluminium shell which, although seemingly bulky, was worth no less than 30–35mph (48–56km/h) to maximum speed!

Record Breaking

Piero Taruffi (in white overalls) with the record-breaking streamlined four-cylinder Gilera on which he clocked over 170mph (273.5km/h) in October 1937. Also visible next to Taruffi are Giuseppe Gilera (in hat) and to his right Ing. Piero Remor.

STREAMLINING PROBLEMS

The original prototype alloy shell encased the rider totally, however, it was discovered that above speeds of 135mph (217km/h), the machine became dangerously unstable and so a revised version was made that did not cover the rider. A fin was also tried, but soon dispensed with. The record breaker had an extended frame over that of the racers, greater spring travel, no front brake and special Pirelli tyres.

THE 1937 RECORDS

The first of the new series of record attempts began in April 1937, being capped by a new Hour Record of 121 miles (195km) (staged over a twenty-eight mile (forty-five kilometre) course on the Bergamo-Brescia *autostrada*). When one realizes that at the Brescia end there was no connecting loop and team members had to physically turn the bike around, it was a superb feat. The previous record had been held by Britain's Jimmy Guthrie on a Norton at 114mph (183km/h).

Then, on 2 October the same year, Taruffi and the Gilera streamliner returned to the same location and broke four more World Records – the five-kilometre at 154.09mph (247.93km/h); the five-mile at 150.15mph (241.591km/h); the ten-kilometre (standing start) at 138mph (222.04km/h), and the ten-mile (standing start) at 132.91mph (213.852km/h). These brought the total of World Records held by Gilera up to no less than eighteen.

Not content with this achievement, Taruffi made more attempts at the end of

Taruffi and Giuseppe Gilera with the 1937 record breaker showing its narrow frontal width.

Record Breaking

Another view of the Gilera streamliner; Luigi Gilera is second from the right.

In April 1939 Taruffi set a new World Record for the hour by covering 127 miles (204km) of the Bergamo-Brescia autostrada. The photograph shows mechanics refuelling Taruffi's machine after the first 62 miles (100km); Ing. Remor looks on.

The record-breaking Gilera squad in 1939 with designer Remor extreme right, and rider Taruffi next to him.

Record Breaking

October, coming within an ace of breaking the outright World Two-Wheel record (held at the time by Eric Fernihough (on a 1000cc Brough Superior V-twin). Taruffi's average speed over the Flying Kilometre was 170.04mph (273.594km/h) which, although fractionally faster than Fernihough's speed of 169.785mph (273.184km/h), did not break the record by a sufficient margin to rank as a record under the sport's governing body's rules. These rules stated that a short distance record was not considered as officially beaten unless the time had been reduced by at least 0.05 seconds – Taruffi failed by a mere 5000th of a second!

The Flying Mile was broken at 169.09mph (272.066km/h), which itself was very close to the outright speed record.

THE 1939 RECORDS

Taking a rest from the record attempts in 1938, Taruffi returned in April 1939, upping the Hour Record to near 127 miles (204km) in the sixty-minute period.

Then came the war. In its immediate aftermath Taruffi found himself ousted from Gilera by designer Piero Remor – ironically Remor himself quit Arcore for MV Agusta over the winter of 1949–1950.

However, Taruffi was not quite out of the picture yet; in the 1940s, with his associates Gianini and Fonzi, he designed and built a tubular, space-framed, twin-boom four-wheeler.

Each boom resembled a torpedo and the driver sat in the nearside section; the conventional steering wheel was exchanged for a series of levers.

THE TARF

Called the Tarf, the twin-boomed car was initially powered by a Guzzi V-twin racing engine; then a Tarf II was built where the driver switched to the offside boom and made use of a 1720cc Maserati four-cylinder car engine.

By 1953 Taruffi had returned to Gilera as a director, and he thus replaced the

On 14 October 1954, Piero Taruffi piloted this strange-looking device; the double-boom machine was named Tarf. At this time Taruffi was the racing director of the Gilera company. His record spree took place at the banked Montlhery circuit, near Paris. Powered by one of the works 500 four-cylinder engines, the Tarf succeeded in breaking some twenty World Records.

Guzzi engine in the original Tarf with one of the latest four-cylinder Gilera engines. Both the standard 500 and a one-off 550cc version were used. With these Taruffi broke a total of eighteen World Records, including covering 125 miles (201km) of the French Montlhéry circuit in the magic hour.

In a career which spanned well over three decades, Taruffi was almost unique in the width of his experiences, racing on both two and four wheels – plus his many record attempts.

TARUFFI'S FINAL YEAR

1957 was his final year; and what an exit it was to be! First, our hero crowned his four-wheel achievements with victory in the tragic last *Mille Miglia*, and later in the year at Monza he collected seven more records in the 350cc category, with his beloved Tarf I now re-engined with the smaller Gilera multi.

Only the legendary John Surtees matched Taruffi for his dual achievements on two and four wheels, and even the great Surtees couldn't lay claim to Taruffi's incredible record-breaking exploits – what a competitor!

THE 1957 RECORDS

As far as Gilera's 'record-breakers' history is concerned, besides Taruffi's efforts, the other big event came after Gilera had announced it was quitting the GP scene in the autumn of 1957. The grand plan was an onslaught on the Standing Start Kilometre and One Hour records for the 125, 175, 350 and 500cc solo, and 350 and 500cc sidecar classes – to be staged at Monza Autodrome.

Initially Gilera decided to concentrate its efforts with Italian riders in the shape of Romolo Ferri (for the smaller classes) and Alfredo Milani for the larger solos and the sidecar.

(Left) *During November 1957, after announcing its retirement from Grand Prix racing, the Gilera factory sent Alfredo Milani (shown here on a 350 four) and lightweight specialist Romolo Ferri to Monza in an attempt to gain new speed records. They were later joined by Bob McIntyre, who set a new Hour Record on the machine Milani is riding here.*

(Right) *Gilera also developed new streamlining (shown here in unpainted form) and tested an outrigger wheel for attempts on sidecar records.*

The attempts started on the morning of 13 November, with the Standing Kilometre length. Although the weather was far from perfect, both Ferri and Milani easily bettered the existing records. But Milani was by no means satisfied with his times and on the following day, in rather better weather conditions, he raised the speeds from 93.74mph (150.83km/h) to 96.28mph (154.92km/h) for the 350cc and from 104.24mph (167.72km/h) to 106.71mph (171.696km/h) for the 500cc.

November 15 should have been the 175cc sidecar 'One Hour' day. There were, however, mechanical problems with both bikes and the attempts had to be postponed.

Saturday 16 November, and there were scenes of great jubilation with the setting of new hour records for the 175cc and 350cc solo categories. In the morning Ferri raised the 175cc figure to 121.663mph (195.756km/h), despite being stopped twice by ignition trouble; whilst in the afternoon Milani's turn came to take the 350cc class record up to 134.133mph (215.82km/h).

Ferri, who also raised the 100km record for his class to 126.189mph (203.038km/h) put in a fastest lap of 131.96mph (212.32km/h). Milani also beat the 100km record at 133.751mph (215.205km/h).

A week later Romolo Ferri added the 100km and One Hour 125cc records to the Gilera bag; while Albino Milani broke the 350 and 500cc sidecar records for the hour and 100km using a 350 four with outrigger attachment.

MCINTYRE'S ONE-HOUR RECORD

But perhaps more important was the news that Bob McIntyre had been invited to assist with the factory's record-breaking attempts.

The fairings of the 1957 record Gileras were all made from 0.5mm thick Elektron sheet. Engines used were the 125 and 175 dohc twins and 350 and 500cc fours – all based on the factory's racing models.

Following the raising of the 175cc 100km record to 130.14mph (209.395km/h) and the 175cc One Hour record to 140.32mph (225.78km/h), the Arcore factory turned its attention to the 350 class and Scotsman Bob McIntyre wrote more history in the Gilera high-speed book on Wednesday 27 November, when he broke the 350cc class 100km and One Hour records. The actual speeds were: 100km – 141.67mph (227.95km/h) and 141.87 miles (228.27km/h) into the hour; both bettering

Albino (brother of Alfredo) Milani setting a new World Record for 350 sidecars with a four-cylinder Gilera fitted with an outrigger wheel.

Record Breaking

Close-up of specially constructed outrigger assembly.

by large margins the records set up only days before with the same machine by Alfredo Milani.

All this was not without some problems. First, before letting McIntyre touch the bike, he was given a full medical to ascertain that everything was alright after his Dutch TT crash (*see* Chapter 3). In addition, his first attempt ended after sixteen minutes due to magneto failure, which entailed the machine being returned to Arcore (only six miles from Monza) for repairs.

Once back at the circuit he succeeded in completing the job – also gaining records in the 500, 750 and 1000cc classes. His fastest lap was at 143.88mph (231.5km/h) and, as *Motor Cycle News* said in their 4 December 1957 issue, 'A terrific speed indeed!'

Romolo Ferri getting the 125 Gilera twin under way during his successful attempt on a number of lightweight World Speed Records at Monza, 13 November 1957.

89

Record Breaking

Many observers have come to realize that McIntyre's breaking of the classic Hour Record was one of the greatest feats in motorcycling history. For a start he rode the 350 Gilera – not the far more powerful 500. He also pushed the old figure – achieved by Ray Amm on the 500cc Norton 'kneeler' in 1953 at Montlhéry – of just over 133mph (214km/h) to over 141mph (227km/h). A speed difference of some 8mph (13km/h) with 150cc less engine capacity. In addition, the rival Moto Guzzi factory failed in its attempt to beat the Gilera records at Monza a couple of weeks later with both their 350cc single and 500cc V-8 machines.

Upon his return, McIntyre said that he had no idea 'that record breaking was so different from racing'. And that he would 'much prefer twelve laps of the TT any day than another record attempt!'

A little-known fact was that the Monza high speed 'saucer' was in fact extremely bumpy – due to its foundations sinking. This was also the main reason why the smaller four was used and is also probably the reason why Guzzi and MV didn't succeed in beating McIntyre's record, as was both factory's original intention.

That One Hour record, achieved on a cold November day at Monza, is a tribute to Bob McIntyre's skill and, together with TT double victory on the Gilera fours, made him the outstanding rider of 1957 – the last year of racing's golden era.

McIntyre's record lasted until Mike Hailwood finally just managed to exceed it on a 500cc MV Agusta at Daytona, prior to the US Grand Prix in February 1964. Even then Hailwood admitted that he was 'engulfed with admiration for McIntyre's efforts'; achieving it as he did on a far bumpier track and with a less powerful machine.

(Left) *Ferri machine with the top section of the streamlining removed. The mechanics of the machine under its streamlined shell were essentially those used on the Grand Prix circuit.*

(Below) *Ferri during his attempt on the 62 mile (100km) distance, at speed on the Monza banking.*

Monza, 13 November 1957, Alfredo Milani setting a new World One Kilometre Sidecar Speed Record with the factory four-cylinder Gilera.

1964 SIDECAR RECORD ATTEMPT

Then, in 1964, came the last records to fall to the legendary four-cylinder Gilera engine, when the Swiss sidecar ace Florian Camathias established the first World Record under a two-year regulation which effectively outlawed the 'wheel on a strut' interpretation of a sidecar.

From 1 January 1962 it had also been obligatory for 132lb (60kg) of ballast to be carried on a sidecar that had to be similar to those used for road racing, complete with platform and nose fairing. This meant that, from the above date, drivers would establish and not break World Records on three wheels.

The record that Camathias therefore *established* was the Standing-Start World Kilometre in a time of 25.32 seconds for an average speed of 88.35mph (142.16km/h). The record counted for the 500, 750 and 1,200cc categories.

But Camathias' real intention had been to better his own (498cc BMW) 500 and 750cc and George Brown's (998cc Vincent) 1,200cc figures. But of course these, having been set under the old rules, were not in fact bettered. In addition, his original intention of going for long distance (six hours and above) had to be abandoned as the banked track at Monza had deteriorated to such an extent that the 500 Gilera outfit was leaping ten to fifteen yards over bumps and was totally unmanageable.

The Camathias Gilera outfit was very low, with a long wheelbase. Both the motorcycle wheels used 16in tyres, the sidecar a 12in cover. The front brake, of the disc type, was pedal operated, while the orthodox twin-cam drum brake on the rear wheel was operated by the right-hand lever on very narrow handlebars. Most of the fuel was carried in a tank on the sidecar platform, with an electric pump, battery driven, to maintain a supply to a small header tank set immediately above the carburettors. Camathias also tried a version of the record breaker in GPs that year, but without any consistent success soon reverted back to his own BMW.

5 Decline and Fall

Although its four-cylinder racing models garnered the press headlines, Gilera's track success was only guaranteed if its standard production models were equally successful. In the immediate post-war period the Arcore company had two models: the Nettuno and Saturno (*see* Chapter 2), both traditional ohv singles with a British heavyweight appearance. However, following then current trends, Gilera introduced a series of totally new lightweights with a modern 'clean' style.

THE 1948 MILAN SHOW

The prototype for this new lineage can be traced back to the Milan Show held from 12–21 November 1948 at the Palazzo dell'Arte. One of the most significant and interesting machines to be seen at the exhibition was a prototype 125 Gilera side-valve four-stroke single.

More correctly, the definition should have been parallel valves. *Motor Cycling* described the layout thus:

A wagon and trailer about to depart the Gilera factory in Arcore with a shipment of motorcycles (125s) in 1950. Note the soft-top Fiat Topolino car in the background.

Decline and Fall

This machine is built symmetrically in that the engine, which has the appearance of being a two-stroke, has a 'T' head, the exhaust valve being arranged centrally in front and the inlet at the back. This of course, necessitates two camshafts. That at the front drives the dynamo on the left-hand side of the machine, and the magneto on the right-hand side. The exhaust pipe is taken between the twin down-tubes, underneath the streamlined casing of the crankcase-gearbox unit, to a silencer from which two tail pipes lead the gases out to each side of the rear wheel.

Other details of this inaugural machine included pressed-steel girder forks and an interesting form of rear suspension embodying a swinging fork and compression springs.

PRODUCTION BEGINS

When production commenced in 1949 much of the chassis of this prototype was employed, but the engine was radically different. Although the principle of parallel valves was retained these were now side-by-side, not front-to-back as on the prototype. The 123.67cc (54 × 54mm) engine featured full unit construction of the gearbox and clutch. There were three speeds, by foot-operation from a lever on the offside – the same side as the kick-starter. Running on a lowly 5.1:1 compression (to suit the low-grade fuel then available), the newcomer could achieve 85kph (53mph). The cylinder was cast iron, but the head was alloy. Carburation was taken care of by a Dell'Orto MA16, with the peak power of 5.5bhp being produced at 5,000rpm. The single-sided brakes were both of 170mm, built onto 19in alloy wheel rims.

Although updated within three years of its launch, the basic layout of the 1949 125 engine (built in both Turismo and Sport versions from 1950 onwards) was good enough to ensure that it was incorporated into a succession of models which were to follow.

Gilera's first production, small capacity, unit construction, four-stroke single, the 1949 125 Turismo. This had a capacity of 123.67cc (54 × 54mm), parallel valves, four-speed gearbox and battery/coil ignition.

Decline and Fall

Gilera's new 125 under construction during 1950. This photograph gives an excellent insight into the production methods of the time. Conveyor-belt assembly had not yet arrived and each machine was assembled on an individual steel or wooden bench.

125 Turismo and Sport

Engine:	Air-cooled, unit construction, four-stroke, ohv, with parallel valves
Bore:	54mm
Stroke:	54mm
Displacement:	123.67cc
Compression ratio:	5.1:1
Maximum power (at crank):	5.5bhp @ 5,000rpm
Lubrication:	Wet sump
Ignition:	Crankshaft-mounted dynamo/battery
Fuel system:	Dell'Orto MA16 carburettor
Primary drive:	Gear
Final drive:	Chain
Gearbox:	Three-speed
Frame:	Tubular steel, using engine as stressed member
Front suspension:	Early bikes blade-type forks; later bikes Telescopic type
Rear suspension:	Early bikes swinging fork with friction damper; later bikes twin shock, swinging arm
Front brake:	Single side drum
Rear brake:	Drum, single-sided, 150mm
Wheels:	19in front and rear
Tyres:	Front 2.00 × 19; Rear 2.50 × 19
Wheelbase:	51in (1,300mm)
Dry weight:	198lb (90kg)
Fuel tank capacity:	2.2imp. gal (10ltr)
Top speed:	53mph (85km/h)

Note:
There was also a larger 152.68cc (60 × 54mm) engine version. Built between 1952 and 1960 this was offered in various guises including: Turismo, Sport, Rossa Super and Rossa Super Sport.

PRACTICAL RATHER THAN EXCITING

During this period, generally Gilera were very similar to rivals such as Moto Guzzi and MV Agusta, the company producing exotic Grand Prix racing machines, but their standard production models were practical rather than exciting. Even so, Gilera still managed to create a stir on occasions, such as the launch of the new B300 twin at the 1953 Milan Show.

THE B300 TWIN

Mechanically it was a fairly conventional machine, with its four-speed gearbox in unit with the engine blending into a smoothly contoured 'power egg' as *Motor Cycling* described it. The separate cylinders, of cast iron, were similarly of clean external appearance, while the light-alloy head had no excrescences at all, the rocker boxes being cast in, and covered by, a finned lid.

Unusual was the choice of capacity, 305.3cc (60 × 54mm). Following a sensible approach, however, Gilera had refused to allow themselves to be cramped by arbitrary capacity limits, and had chosen instead to design according to the requirements of the job in hand. In any case, this had the added advantage of ensuring they were able to employ several components from the well-developed 150 model (introduced a year before in Turismo and Sport guises – 152.68cc – using the same 60 × 54mm bore and stroke measurements).

In general appearance, the twin followed the smaller models with its telescopic forks, twin shocks, swinging arm rear suspension and full-width polished alloy brake hubs. The wheel rims were of light alloy and so, too, were the Gilera-built silencers. Although lacking any real power (12.5bhp at 5,800rpm on the original production model), the B300 did have the advantage of being exceptionally smooth, easy to start with good flexibility.

(Above) *The sensation of the thirty-first Milan Show, held at the end of 1953, was the launch of the new Gilera B300 ohv vertical twin. Actual capacity was 305.3cc (60 × 54mm) – which just happened to be the same as the company's existing 150 model.*

(Below) *English journalist John Thorpe testing one of the newly released Gilera B300 twins in Italy during November 1953, prior to the beginning of the Milan Show that year. He was one of the very first non-Italians to do so.*

Decline and Fall

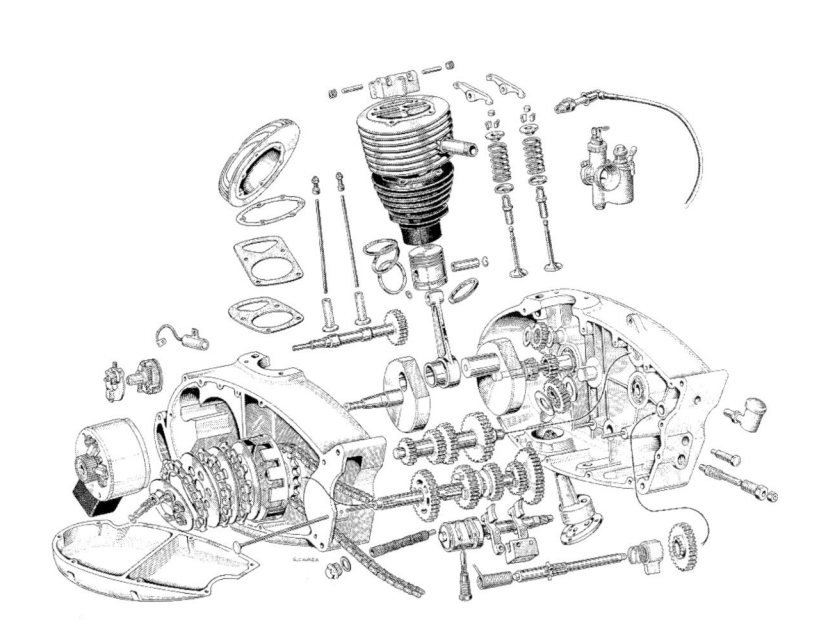

From the 125 came the 150. This first appeared in 1952 and continued until 1960. The larger engine capacity was achieved by increasing the bore size from 54 to 60mm; the stroke remained unchanged.

The 300 Extra (and also 250 Export) were both launched at the Milan Show in November 1955, going on sale in early 1956. The most notable visual change compared to the standard model was the fitting of a dual seat. Later the Extra gained Silentium-made silencers in place of the Gilera-made alloy-bodied originals.

When *Motor Cycling* tested an example in their 9 December 1954 issue, they recorded a maximum speed of 70mph (113km/h), hardly breathtaking, but this was a tourer, not a sportster. Much of the engine also followed features pioneered on the 125/150 singles – such as parallel valves, coil valve springs, wet multi-plate clutch, built-up crankshaft, four-ring piston and dynamo lighting. The cylinders were inclined forwards ten degrees from the vertical.

(Above) *Developed from the earlier Rondine design, the water-cooled, supercharged, four-cylinder Gilera was the fastest GP bike of its era – and European Champion in 1939.*

The 1951 Gilera Saturno single. Its 498.76cc (84 × 90mm) ohv engine produced 22bhp at 5,000rpm. Suitable for a wide range of roles, the Saturno was Italy's BSA Gold Star.

(Below) *Dustbin faired, four-cylinder Gilera, of the type used by Bob McIntyre to win the double (350 and 500cc) Isle of Man Jubilee TTs in 1957.*

(Above) *Fully faired 125 twin GP bike at Assen, Centennial TT, May 1998.*

(Left) *Engine from the 1956–1957 124.656cc (40 × 49.6mm) dohc Grand Prix twin-cylinder model.*

(Below) *The 1957 175 Formula Two racer. This design was intended for use in events such as the Milano–Taranto and Giro d'Italia (Tour of Italy) long-distance events.*

The 250 Arizona of 1985 was a pukka enduro machine. Powered by a single-cylinder, water-cooled engine, it featured reed valve induction, six speeds, long-travel suspension, map case and hand guards.

Taken over by Piaggio in 1969, Gilera became a major force on the Italian domestic scene during the 1970s and 1980s. This is one of the liquid-cooled, sporting RV125 models of the mid-1980s.

Engine from the KK-KZ125 series. There was even a one-model race championship for these fleet little sportsters.

Former Gilera works rider and three-time world 500cc champion Geoff Duke, with one of the Arcore fours before setting out on a TT parade lap during the early 1980s.

A view of production facilities at the Arcore factory in 1987.

In 1988, Gilera broke a number of speed records for standard production 125cc models using the new MX1, which had been launched at the Milan Show the previous November.

(Left) Thanks to a Japanese company, Gilera built the Nuovo Saturno, using an engine based on the dohc, four-valve Dakota single-cylinder trail bike, which itself had been launched at the 1985 Milan Show.

(Below) This neat little Paris–Dakar styled, XR2, street-enduro bike was powered by an engine based on the MX1 racer replica.

The XRT was a development of the earlier Dakota, but used a torquey 569cc (99 × 74mm) engine, which, coupled to a relatively light weight of 330.5lb (150kg), gave excellent performance, circa 1988.

At the beginning of the 1990s Gilera built small batches of the racing-only Nuovo Saturno Piuma *(Puma). This was intended for use in the popular Supermono (singles) racing series, but it proved prone to crankcase failure and excess vibration.*

Gilera were much more successful off-road, with the factory RC600R enduro machines winning the 600cc class of the Paris–Dakar Rally on more than one occasion.

A 1993 Nordwest 600. Although offering ultra-modern styling, the bike was also an excellent motorcycle with adequate performance and superb on-road ability. Some were even raced successfully.

Aprilia had its Extrema, Cagiva its Mito – and Gilera the GFR. These hyper-performance 125s could all achieve amazing performances. This is the SP version with well over 100mph (160km/h) available on the track.

During 1992 and 1993, Gilera returned to GP racing with a brand new twin-cylinder two-stroke. Designed by Frederico Martini, it struggled to compete with the factory Aprilias and Hondas. But it was still a mightily attractive machine.

Road Test B300 Twin – Mick Walker

When Gilera announced, in 1953, that it was to offer a parallel twin, it created considerable attention, both in the Italian press and overseas; this was, after all, an engine type which had gained a considerable post-war following in Great Britain, if not elsewhere in Europe. Certainly, in Italy, the trend was for smaller capacity machines with a single lung.

At first glance, the B300 seemed a fairly straightforward design, with the engine following conventional Gilera lines of the time (the Saturno excluded) with the four-speed gearbox in unit with the engine. The separate cylinders of the 305.3cc (60 × 54mm) power unit were made of cast iron, with separate aluminium cylinder heads – and above these latter components was a pair of what are best described as finned lids, covering the valve gear.

Why a '300' you might ask, instead of a 250 or 350? Well, the answer was simply that the engine was a 'double-up' of the existing and well-proven 150 (152cc actually). The bottom end, as on the small capacity Gileras of the same era, was almost egg-shaped, but to accommodate the extra cubes of output everything was uprated. For example, the design team saw fit to equip the B300 with a duplex chain rather than the helical gears that the smaller singles used for primary drive. The transmission gears themselves were wider and thus stronger, the kick-start was on the right not the left, and the built-up crankshaft not only had a massive central section missing on the singles, but plain metal big-end bearings. Both connecting rods ran to a single long crankpin that passed through not only the massive central flywheel, but the conventional outer crank flywheels of the type found on the singles.

As on the small capacity ohv single there were parallel valves – and, as on the 150, oversquare bore and stroke dimensions – the other single usually being equal or long-stroke. Another feature was the rear-facing position for the spark plugs.

The electrical equipment, as on other Italian motorcycles of the period, was of six volts. The B300 featured a crankshaft-mounted dynamo, with a battery and twin contact breakers/condensers (one for each per cylinder).

The majority of the cycle parts were based on those used by the singles, with a tubular steel frame that sported twin front downtubes, but no bottom rails – the engine acting as a stressed member.

Both the oil-damped telescopic forks and twin rear shock absorbers were fully enclosed by steel/alloy covers. Final drive was by simplex chain on the offside of the machine.

The first series machines had several differences to the later (post-1955) model that I tested here. These included a single sprung saddle and parcel rack on rear mudguard (dual seat), smaller 160mm diameter front brake (against 180mm), a 3.3imp. gal (15ltr) as opposed to 4.18imp. gal (19ltr) fuel tank and aluminium Gilera-made silencers (Silentium-made chromed steel).

The B300 was the first Gilera production roadster to use 18in wheels. Again, the first batches of machines employed welled aluminium rims, later bikes having cheaper chromed steel components.

Out on the Road

The first thing that struck me about the B300 was how compact the machine was for a twin – the second was how docile it felt. The smoothness, ease of starting and light weight were all equally noticeable.

The smoothness of the power delivery can partly be attributed to the small size of the valves, the ports and the use of a single 22mm Dell'Orto carb. Yes, this was most definitely a touring, rather than a sporting, motorcycle.

The quietness of the engine, the conventional handlebars (in a time when clip-ons were to be

Decline and Fall

> **Road Test B300 Twin – Mick Walker** *(continued)*
>
> found on many Italian bikes) and the deeply valanced mudguarding, all spoke of comfort over a racy style. In many ways the B300 was a pleasure to ride – except for an over-hard seat (the dual seat was fitted), insufficient ground clearance (but probably entirely suitable if used in its intended touring/commuter role) and brakes that had adequate, rather than brick-wall-stopping, power. In fact the 160mm front (albeit full width) of the original 1953–1954 version would have most definitely been the design's biggest single failing.
>
> Performance, or the lack of it, was another problem. 15bhp at 6,800rpm was never going to set the world on fire. And, even though the Gilera twin was a true lightweight, its acceleration and maximum speed (the latter about 75mph (120km/h)) were not too impressive.
>
> However, this was only half the story, as the engine's inherent smoothness and its ability to cruise all day at around 70mph (112.5km/h) went some way to offsetting the lack of outright 'go'. Another plus was just how easy it was to start. A simple prod of the kick-start lever would usually suffice – you could do it by hand, it was that simple! Try doing the same on a single …
>
> The remainder of the bike was very much 1950s–1960s Italian lightweight fare which meant beautiful aluminium castings, but lacklustre paintwork and chrome. Nowadays, of course, modern restoration techniques can cure these when-new weaknesses. The engine appeared almost over-engineered and lightly stressed. This just left the electrics and what can anyone truthfully say of the older Italian bike electrics? Again, with modern technology, even this area could probably be sorted out.
>
> So, there you have it – the B300 twin was a pretty rare animal – a 1950s Italian ohv-engined machine with more than one cylinder, a smooth, if rather than tame performer but with its own unique character. However, in the end it somehow left me wanting more. If only it could have been a slightly bigger overall bike, with a larger displacement (maybe 500cc minimum) and with a five-speed gearbox. However, of course it should be remembered that it was created when the war had only recently ended; austerity still ruled and the early 1950s was a time when it was a seller's, rather than a buyer's, market. Taken into consideration that makes the Gilera twin a pretty good effort.

B300 (1953–1969)

Engine:	Air-cooled, unit construction, twin cylinder, ohv, with parallel valves
Bore:	60mm
Stroke:	54mm
Displacement:	305.3cc
Compression ratio:	6.5:1
Maximum power (at crank):	15bhp @ 6,800rpm
Lubrication:	Wet sump
Ignition:	Crankshaft-mounted dynamo/battery
Fuel system:	Single Dell'Orto MB22B carburettor
Primary drive:	Duplex chain
Final drive:	Chain
Gearbox:	Four-speed
Frame:	Tubular steel, with twin front down-tubes
Front suspension:	Telescopic fork, enclosed type
Rear suspension:	Twin shock, with swinging arm

Decline and Fall

B300 (1953–1969)

Front brake:	Drum; full-width aluminium
Rear brake:	Drum
Wheels:	18in front and rear
Tyres:	Front 3.00 × 18; Rear 3.25 × 18
Wheelbase:	53in (1,350mm)
Dry weight:	330.5lb (150kg)
Fuel tank capacity:	1953–1954 3.3imp. gal(15ltr); 1955 onwards 4.18imp. gal (19ltr)
Top speed:	77.5mph (125km/h)

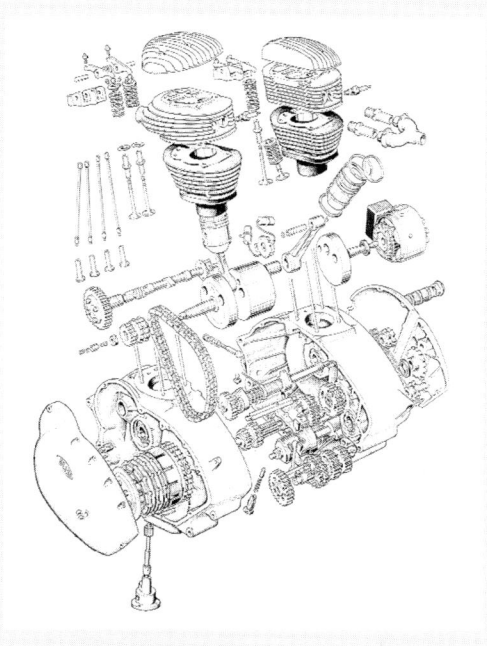

(Top left) *Original B300 model with single sprung saddle, parcel carrier, small front brake, 15ltr tank and aluminium silencers.*
(Middle left) *The compact nature of the 305.3cc (60 × 54mm) unit construction engine is evident in this view. Note the duplex front-frame downtubes.*
(Top right) *Exploded view of the B300 engine assembly, showing details such as the duplex primary chain, parallel valve gear, four-ring pistons, crankshaft-mounted dynamo, iron cylinders and alloy heads.*
(Bottom left) *A late-type B300 twin of the type tested by the author. Note the larger diameter front brake, Silentium chrome-steel silencers and 19ltr fuel tank.*

Decline and Fall

THE 175 IS INTRODUCED

Nothing else of real interest on the production line appeared until the 1956 model year, but at least when it finally appeared it was destined to become a mainstay of Gilera production until the beginning of the 1960s. This was the '175' single, which was built in a number of guises including the Sport, Super Sport, Regolarita (ISDT-type events) and Cross (motocross).

With an actual capacity of 172.47cc, the bore and stroke dimensions were almost square at 60 × 61mm. The valve layout was once again of the parallel type. The spark plug was mounted on the offside of the cylinder head, whereas the B300 twin had its plugs set at the rear of the cylinder head just outboard of the carb inlet stub.

In a 1958 road test *Motor Cycling* referred to the 175 as 'an amazingly economical, high-performance race-bred sports lightweight'.

MOTOR IMPORTS

Brought into Britain by Motor Imports of London (the trade arm of Pride & Clarke), the 175 retailed at £199.

Over the measured Flying Quarter Mile, the machine averaged 65mph (105km/h), whilst at the other end of the spectrum 128mpg (2.2/100km) was achieved at 30mph (48km/h), falling to 96mpg (2.95/100km) at 50mph (80.5km/h) – but still highly praiseworthy.

Compared to other Italian bikes of the period the electrics received praise, the *Motor Cycling* tester commenting:

> An excellent driving light is an outstanding feature, its beam shape being a horizontal strip spreading to both sides of the road with an intense whiteness. Moreover, there was a most courteous cut-off when the main beam was dipped. The electric horn, operated from a control cluster on the left bar, was adequate for most purposes and better than usual.

Making its début in 1956, the 175 Sport ultimately proved one of Gilera's most popular models. With a capacity of 172.47cc (60 × 61mm) it followed the general design of the earlier 125/150 series. In fact, in appearance the engine was virtually identical to the 150. But a give-away was the number of fins on the cylinder barrel – the 150 had five, the 175 six.

Decline and Fall

For the 1957 season Gilera introduced the Rossa Extra, essentially a de luxe version of their popular 175 Sport. Besides a change in colour scheme, the newcomer had several other changes. For a start the engine was revised with the drive to the rear wheel being transferred from the nearside to the offside. Next, a full-width alloy rear brake hub replaced the earlier single-sided assembly. But the most noticeable innovation was the fitting of a piggy-back dual Silentium silencer system on the nearside of the bike. This again contrasted with the Sport which had a single unit on the offside.

The test ended:

> Certain details apart – a better conceived centre stand and a more durable finish would have received our blessing – the Gilera is a likeable machine, clearly owing much both to its Italian heritage and to its factory's racing experience. It has a dual appeal: in its touring trim it would be a fast and very economical lightweight and, in its semi-sports form, as tested, it would be quite at home when competing, without lights, silencer and centre stand, in a Clubman's-type event for the appropriate capacity.

It is worth noting that Gilera did update its various models during their production life, although this was often merely cosmetic rather than mechanical. However, the long running B300 (1953 through to 1969) did undergo some technical changes during its existence, typified from 1956 when the engine was uprated to give 15bhp and an additional 500 extra revolutions. There was also a 250 Export variant; but this was not built in large numbers. The '300' was sold throughout much of the 1960s across the Atlantic in the United States.

THE JUBILEE SERIES

To celebrate its Golden Jubilee, 1959 saw the Arcore factory introduce a new lightweight motorcycle appropriately named the *Giubeleo* (Jubilee). There were two versions, the Normal and the Extra, the latter with a higher specification. At first glance the Giubeleo might have appeared to be simply a smaller version of the existing ohv singles that Gilera had been building for the previous ten years. In truth its engine was somewhat different. In particular, the camshaft was mounted on the nearside of the crankcase and in front of the cylinder. The contact breaker, easily accessible from the outside, was keyed to the gear-driven camshaft, while the nearside wall of the cylinder contained a tunnel for the pushrods. However, the parallel valves were retained. The diameters of these valves were: inlet 19mm, exhaust 17mm. Unlike the B300 twin, for example, which had a duplex primary chain, the newcomer employed helical gears and these also actuated the oil pump for the wet-sump lubrication system.

Decline and Fall

175 Sport/Super Sport (1956–1960)	
Engine:	Air-cooled, unit construction, single-cylinder, four-stroke, ohv with parallel valves
Bore:	60mm
Stroke:	61mm
Displacement:	172.47cc
Compression ratio:	6.5:1
Maximum power (at crank):	Sport 7.5 @ 6,000rpm; Super Sport 9.1 @ 6,000rpm
Lubrication:	Wet sump
Ignition:	Crankshaft-mounted dynamo/battery
Fuel system:	Sport: Dell'Orto MA18B carburettor; Super Sport: Dell'Orto MA22B carburettor
Primary drive:	Gears
Final drive:	Chain
Gearbox:	Four-speed
Frame:	Tubular steel, twin front down-tubes
Front suspension:	Telescopic fork, enclosed type
Rear suspension:	Twin shock, with swinging arm
Front brake:	Drum; full-width, aluminium hub
Rear brake:	Drum; full-width, aluminium hub
Wheels:	19in front and rear
Tyres:	Front 2.00 × 19; Rear 2.50 × 19
Wheelbase:	51in (1,300mm)
Dry weight:	Sport 229lb (104kg); Super Sport 242.5lb (110kg)
Fuel tank capacity:	2.85imp. gal (13ltr)
Top speed:	Sport 71mph (115km/h); Super Sport 73mph (118km/h)

To celebrate fifty years since Giuseppe Gilera had built his original 317cc single back in 1909, the Arcore Factory built various Giubileo (Jubilee) models in 1959. One of these was the 124 Giubileo Extra, which was almost an enlarged version of the 98 Giubileo launched earlier that year.

The square 50 × 50mm (98.173cc) engine developed 6bhp at 7,000rpm and with a 7.8:1 compression ratio offered maximum torque at 4,700rpm. Cycle parts, including the duplex frame, telescopic forks, swinging arm rear suspension and full-width hubs followed conventional Gilera lines, as did the emergency electrical system, which could be employed if the battery was flat. Like the majority of Gilera's production roadsters of the period, the 98 Giubeleo was not in the sporting category and therefore outright speed didn't feature as a priority. A shade over 50mph (80km/h) was possible, but it would hold this figure for many miles without stress. Like the other singles, Gilera built a Regolarita version, some of which were entered by the factory in long-distance trials, most notably the ISDT, with considerable success.

98cc DETAILS

An interesting feature of the 98 street bike was its gearbox; this featured a 'low' bottom (final ratio was 29.150:1) which proved very useful. Two rear sprockets were offered (54 and 55 teeth), the latter specified for intensive mountain terrain, when the bottom gear ratio was 29.688:1.

Other features of the small machine's specification included: Silentium silencer, 7.8:1 compression four-ring piston, 6V ignition with 32W flywheel-mounted generator, 18mm (some early bikes had 16mm) Dell'Orto carb, four-speed gearbox, 17in tyres, 123mm diameter full-width drum brakes and a dry weight of 227lb (103kg).

ISDT SUCCESS

Three specially prepared 98 Giubeleo Regolarita models secured an impressive trio of gold medals in the 1960 ISDT in Austria when, together with 125 versions, a Gilera team won the Silver Vase for Italy.

The 175 Jubilee was launched in 1960, with production continuing until 1966. Features included 10bhp at 8,000rpm, 7:1 compression ratio, 158mm diameter full-width alloy brake hubs, four speeds and 17in wheels. Maximum speed was 69mph (111km/h).

Decline and Fall

Giubeleo (Jubilee) 98 (1959–1970)

Engine:	Air-cooled, unit construction, single-cylinder, ohv, with parallel valves
Bore:	50mm
Stroke:	50mm
Displacement:	98.173cc
Compression ratio:	7.8:1
Maximum power (at crank):	5.8bhp @ 7,000rpm
Lubrication:	Wet sump
Ignition:	Flywheel magneto 6v, 32w
Fuel system:	Dell'Orto DA18mm carburettor
Primary drive:	Gears
Final drive:	Chain
Gearbox:	Four-speed
Frame:	Tubular steel, engine as stressed member
Front suspension:	Telescopic fork, enclosed type
Rear suspension:	Twin shock, swinging arm
Front brake:	Drum; full-width, aluminium SLS
Rear brake:	Drum; full-width, aluminium SLS
Wheels:	17in front and rear
Tyres:	Front 2.50 × 17; Rear 2.75 × 17
Wheelbase:	49in (1,250mm)
Dry weight:	Sport 227lb (103kg)
Fuel tank capacity:	2.4imp. gal (11ltr)
Top speed:	Sport 49mph (79km/h)

A NEW 125 ARRIVES

The Jubilee theme had quickly spread to a new 125 (123.15cc, 60 × 56mm) later in 1959. Basically this revised 125 ran from 1959 through to 1970, its major change along the way being the introduction of a five-speed gearbox in 1965. Factory definition for the four- and five-speed gearboxes were 4v and 5v, respectively. There was also a 150 version with 158.33cc (60 × 56mm). The 150 was only built in the four-speed guise.

The final Jubilee variation was the 202 which appeared in 1966. Essentially the engine owed much to the earlier 175, but with a five-speed gearbox and a larger cylinder bore of 65mm; the stroke remained unchanged.

THE SCOOTER PROJECT

Earlier in 1962 the G50 scooter made its début. This was soon followed by a larger version, the G80. Both featured remarkably economical pushrod engines – again featuring the Gilera hallmark of parallel valves. The smaller unit had a capacity of 49.9cc (38 × 44mm) and the G80, 76.44cc (44 × 44mm). Both employed a twist-grip operated three-speed gearbox and primary drive was by helical gears. Although both versions were as good as anything in their class, they didn't sell in the numbers required. Quite simply, by the time they were introduced the scooter boom was in rapid decline.

Decline and Fall

Giubeleo Super 202 (1966–1970)	
Engine:	Air-cooled, unit construction, single-cylinder, ohv, with parallel valves
Bore:	65mm
Stroke:	61mm
Displacement:	202.4cc
Compression ratio:	7:1
Maximum power (at crank):	11bhp @ 6,500rpm
Lubrication:	Wet sump
Ignition:	Crankshaft-mounted dynamo/battery
Fuel system:	Dell'Orto UBF 22BS carburettor
Primary drive:	Gears
Final drive:	Chain
Gearbox:	Five-speed
Frame:	Tubular steel, engine as stressed member
Front suspension:	Telescopic fork, enclosed type
Rear suspension:	Twin shock, swinging arm
Front brake:	Drum; full-width, aluminium SLS
Rear brake:	Drum; full-width, aluminium SLS
Wheels:	17in front and rear
Tyres:	Front 2.75 × 17; Rear 3.00 × 17
Wheelbase:	50in (1,280mm)
Dry weight:	260lb (118kg)
Fuel tank capacity:	3.3imp. gal (15ltr)
Top speed:	71.5mph (115km/h)

Gilera built a couple of scooters in 1962/63. These were the G50 (shown) with 49.9cc (38 × 44mm) and G80 with 76.44cc (46 × 46mm) ohv four-stroke engines and three-speed gearboxes. Finally phased out in 1966, neither succeeded in making much contribution to the company's balance sheet.

G50/G80 Scooter (1962–1966)

Engine:	Air-cooled (fan-assisted) unit construction, ohv with horizontal single cylinder
Bore:	38mm (46mm)
Stroke:	44mm (46mm)
Displacement:	49.9cc (76.44cc)
Compression ratio:	8:1 (7.8:1)
Maximum power (at crank):	1.48bhp @ 4,800rpm (3.65bhp @ 6,000rpm)
Lubrication:	Wet sump
Ignition:	Flywheel magneto 6v/18w (6v/28w)
Fuel system:	Dell'Orto SH14 12/2 carburettor
Primary drive:	Gears
Final drive:	Direct
Gearbox:	Three-speed
Frame:	Pressed steel
Front suspension:	Trailing link
Rear suspension:	Swinging fork
Front brake:	Drum
Rear brake:	Drum
Wheels:	10in front and rear
Tyres:	3.00 × 10 front and rear
Wheelbase:	44in (1,120mm)
Dry weight:	G50 143lb (65kg); G80 145.5lb (66kg)
Fuel tank capacity:	0.88imp. gal (4ltr)
Top speed:	G50 25mph (40km/h); G80 45mph (73km/h)

Note:
Differences for G80 in brackets.

AND A MOPED TOO

Besides the scooter project Gilera brought out the Gilly moped in 1964. This was powered by a 47.63cc (38 × 42mm) piston ported two-stroke engine with three gears, operated by a twist-grip control.

THE CADET ULTRA-LIGHT

Two years later and the Cadet 50 ultra-light motorcycle made its début.

The Cadet was equipped with an ohv four-stroke engine of 47.63cc (38 × 42mm) featuring a vertical cylinder. None of the scooters, mopeds or ultra-lightweights achieved any real showroom success and it was largely left to the by-now aging singles, together with the long-running 300 twin, to prop up the rapidly ailing fortunes of the Arcore factory.

GILERA IN THE USA

Seven Gilera models were offered for sale in the USA during 1964 by C & N Enterprises of Long Beach, California, including the 98, 124 and 175 singles, plus the 300 twin. But even though the importer took full-page advertisements in *Cycle World* they never sold Stateside in any real quantities, if for no other reason

Gilly Moped (1964–1966)

Engine:	Air-cooled, unit construction, two-stroke single-cylinder, with piston-port induction
Bore:	38mm
Stroke:	42mm
Displacement:	47.63cc
Compression ratio:	6.95:1
Maximum power (at crank):	1.06bhp @ 4,200rpm
Lubrication:	Petroil
Ignition:	Flywheel magneto, 6v/18w
Fuel system:	Dell'Orto carburettor
Primary drive:	N/A
Final drive:	Chain
Gearbox:	N/A
Frame:	Pressed steel
Front suspension:	Undamped telescopic type
Rear suspension:	Twin shock, pressed steel swinging arm
Front brake:	Drum
Rear brake:	Drum
Wheels:	19in front and rear
Tyres:	2.00 × 19 front and rear
Wheelbase:	44.5in (1,130mm)
Dry weight:	103.5lb (47kg)
Fuel tank capacity:	0.88imp. gal (4ltr)
Top speed:	23mph (37km/h)

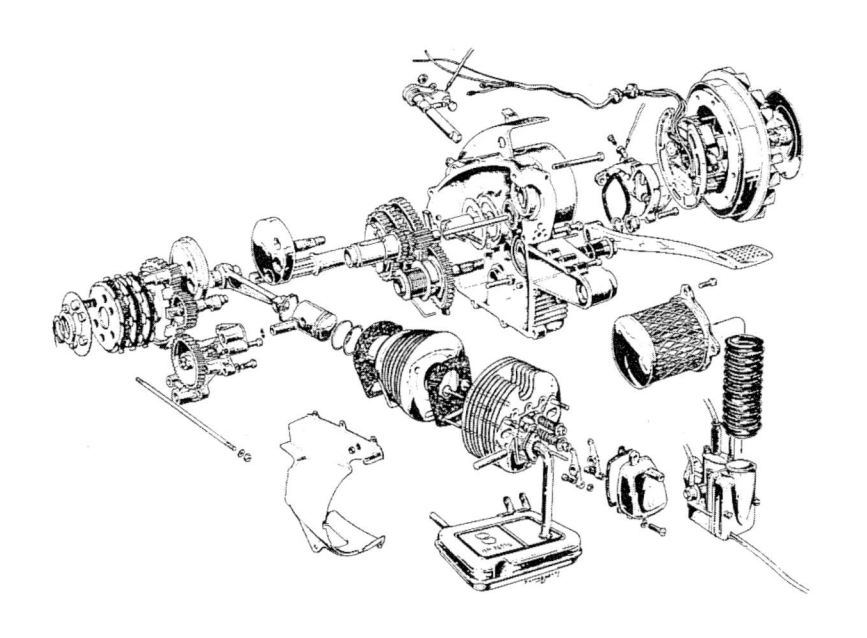

Exploded view of the G50/G80 scooter engine. Note the flywheel magneto with integral fan blades for engine cooling, and flat car-type exhaust box.

Decline and Fall

(Left) *Engine of the 1964 124* Sei Giorni Speciale *(Six Days Special). By the time this arrived Gilera had changed the original bore and stroke measurements of their 125 pushrod engine to 56 × 50mm, giving 123.15cc. Evident in this view is the pushrod tunnel cast, integral with the cylinder barrel, large breather canister, points housing, heavily finned sump and twin front-frame tubes.*

(Below) *Offside view of the 1964 Six Days Special. This is a British registered bike imported by Gilera's UK agents, Motor Imports (trade division of dealers Pride & Clarke).*

than Honda and the other Japanese manufacturers had already begun their massive sales drive into the North American market.

GILERA STRUGGLES TO SURVIVE

With an ever worsening balance sheet, the factory back in Arcore, still with Giuseppe Gilera at the helm, struggled on. No doubt they did their best, as the ultra-lightweight models described above testify. In a market suffering from rapidly falling sales what could the company do when saddled with a massive plant, which the Gilera facility had become over the years? One answer seemed to be to follow rival Moto Guzzi's lead – military and police contracts. So when the Italian government issued tenders for a new high-performance

Decline and Fall

GILERA

with bloodlines of International Champions; the heritage of greatness, comes seven new 1964 models that are unequalled in quality.

The 98cc single cylinder four-stroke o.h.v. engine provides complete efficiency throughout the performance range as only a four-stroke can. All models also feature a four-speed transmission, telescopic forks with hydraulic dampers, hydraulic rear shocks and many other "big bike" features. Sport and Standard models available in Red and White, or Black and White. Accept no compromise, buy GILERA.

— Since 1909 —

The **124cc** over-square engine is a revelation to ride; quick, powerful, and economical. All this and Italian beauty, too! Standard or Sport.

The **175cc**, in true GILERA tradition, offers another superb machine, unequalled in its class, and a genuine pleasure to own. Comes in Red and Black.

A **300cc** o.h.v. twin for that "extra smooth" riding pleasure in, or out of town. This fine motorcycle will satisfy even the most discriminating!

It costs much more to produce this top quality four-stroke motorcycle line — will you pay just a little more for the very best?

"Owning a GILERA is the epitome of distinction!"

Mr. Dealer: Here is a top quality Italian motorcycle line you don't have to apologize for, before or after a sale. The GILERA line will compliment your present inventory and when the "4's" are available every GILERA owner will be a hot prospect, as will every other motorcycle owner.

SEE THE COMPLETE GILERA LINE
AT OUR LONG BEACH LOCATION
Importer/Distributor — Inquiries Invited
C & N ENTERPRISES
750 Long Beach Boulevard
Long Beach, California
PHONE: HE 5-3486

The American Gilera importer, C & N Enterprises of Long Beach, California advertised the Italian marque throughout the States, including this full page advertisement in the December 1963 issue of Cycle World. *Sales never matched the advertising campaign.*

Decline and Fall

motorcycle to replace its ageing Guzzi Falcone 500 singles, Gilera entered the race, together with Guzzi and Ducati. The Gilera design was a 483.02cc (71 × 61mm) with a chain-driven single overhead camshaft and parallel valves. Known as the B50 5V (denoting five-speeds), it was Gilera's first 500 since the famed Saturno.

Initial work on the project began in early 1966 and when the prototype was completed a year later it produced 32bhp at 7,000rpm and appeared a robust and neatly styled machine. Right from the start Guzzi, however, had the upper hand with their 700 V7, and both the Ducati (an 800 vertical twin) and Gilera's offerings were soon sidelined from the contest.

THE B50 TWIN

Having spent considerable time and expense on the big two, Gilera produced a 'civilized' version, the B50. This was very similar to the original, but without the military drab finish and a few styling changes, plus more power: 40bhp at 7,500rpm; the dry weight was 180kg (397lb). Both machines were the work of the same designer as the Saturno, the much travelled Ing. Giuseppe Salmaggi.

THREE-CYLINDER 750 PROTOTYPE

Salmaggi also designed a dohc twin (in both 350 and 500 forms) and a three-cylinder 750, the latter (which was built) had lines very similar to the, then current, Honda CB750 four. Unfortunately, by the time the B50 and its brothers were nearing the production stage in late 1968, Gilera was already in deep financial difficulties and so all these projects were shelved.

As for Comm. Giuseppe Gilera himself, worn down by the factory's difficulties and still mourning the loss of his beloved son, he had finally given up the will to run the enterprise which he had worked so hard to create.

The first prototype of Gilera's B50 ohv twin as it appeared in the autumn of 1967. Originally built to police/military requirements, the Arcore company decided to press on and develop the machine for general use, after the design was beaten by Guzzi's V7 for government contracts. Unfortunately, it was never destined to reach series production and was scrapped after the Piaggio take-over in 1969.

Decline and Fall

Another Gilera project of the late 1960s was the design of a completely new modular series of engines – a 350/500 twin and a 750cc triple – the 500 twin being shown here. These were of modern concept with electric start, disc front brake and Honda CB750 type-styling, but, again, they never reached the production stage.

A RECEIVER IS APPOINTED

Finally, in November 1968, a receiver was appointed at the Arcore plant, where 280 of the 550 workers were already working on short time.

PIAGGIO SAVES THE DAY

A few months later the giant Piaggio organization gained control (after paying a reported £2 million). As for Gilera himself, his retirement was to be relatively short-lived, and on 21 November 1971, just before his 82nd birthday, the great man passed away.

However, thanks to the Piaggio rescue, his name at least lived on for a new generation as Gilera had, by now, become Italy's oldest surviving motorcycle marque.

6 Piaggio

The Piaggio company was founded in 1884 in Genoa, making wood-working machinery for the ship building industry. In 1901 it turned to railway rolling stock and then to aircraft. An aviation plant was built at Pisa (of Leaning Tower fame) before World War One and a car manufacturing facility was acquired at Pontedera in 1924, which was extended for aero-engine and aircraft production to such an extent that, by 1939, Piaggio had 10,000 workers and had attained the leading role in the Italian aviation industry. Its most famous wartime type was the P108 – a four-engined heavy bomber, and the only such type manufactured in Italy during the hostilities.

THE VESPA ARRIVES

By 1944 nothing was left but ruined buildings and a workforce trustfully awaiting their next meal ticket. With the design of a small, two-stroke auxiliary engine and hardly any machine tools, Doct. Enrico Piaggio decided that what he could best contribute to a practically derelict country was basic transportation, so he summoned his management team and instructed them to produce something that would meet the bill. The result of this was the design of a two-stroke vehicle embodying the latest technology of motorcycle, automotive and aircraft engineering – the now legendary Vespa scooter.

Aviation

In a bizarre twist of fate, it was aviation that linked Gilera and Piaggio long before the latter took over the former in 1969.

This came in August 1923 when, following a government directive, Piaggio absorbed the *Pegna-Bonmartini Costruzioni Navali-Aeronautiche* organization. And, as covered in Chapter 1, it was Count Giovanni Bonmartini who, together with Carlo Gianini and Piero Remor, had been responsible for the GRB/Rondine projects, which had ultimately become the Gilera four!

Piaggio themselves had entered the aircraft industry during the First World War, building Caproni bombers.

With factories at Sesti Ponete (Genoa), Finale-Ligure, Pisa and Pontedera, British Bristol and French Gnome-Rhone engines were built under licence.

Besides the several Piaggio-Pegna aircraft, Piaggio also entered into other joint ventures, including the construction of a number of German Dornier Wal flying boats – again under licence.

In 1930, Piaggio, together with designer Ing. Corradino d'Ascanio, built the first d'Ascanio *Elicottero* (helicopter) which flew successfully on 8 October that year. Further helicopter projects were conceived over the following decade. Another d'Ascanio-Piaggio venture was to result in the commercially successful production of variable-pitch propellers.

Aviation

In 1929 Ing. Pegna designed the P7 Schneider Trophy racing seaplane. This revolutionary machine employed a pair of tiny hydrovanes to eliminate the drag imposed by conventional floats; at first the aircraft rested in the water and was driven by a small, two-bladed water-screw below the rudder, but, as the speed increased, the vanes lifted the aircraft out of the water to a point where the airscrew could be engaged. With a planned maximum speed of 373mph (600km/h), the P7 would certainly have been competitive. However, the water-screw clutch slip proved to be its Achilles heel and so the more conventional designs by Fiat and Macchi represented Italy instead.

In 1936 Ing. Giovanni Casiraghi (formally with the American Waco Aircraft company) was appointed chief designer. The most well-known models to appear over the next few years were the P32 twin-engined bomber and the P108 four-engined bomber-transport, the latter being the only Italian four-engined bomber to see wartime service.

The final wartime Piaggio design that actually flew was the P119 single-seat fighter and, like most Piaggio aviation projects, was of extremely original concept. Well before the outbreak of hostilities Piaggio had been studying the possibility of employing a radial engine within the central section of the fuselage (in other words, buried), with the propeller being driven by an extension shaft. Piaggio's engineering team realized that this set-up would not only solve Italy's lack of a high performance liquid-cooled inline engine (such as the British Rolls-Royce Merlin and German Daimler Benz DB series) but also give a central weight distribution, excellent aerodynamics and a concentration of fire power by closer grouping of the guns.

The first P119 flight came on 19 December 1942, and subsequent tests revealed a maximum speed of 398mph (640.5km/h), a service ceiling of 41,340ft (12,600m) and a range of 940 miles (1,513km). But no production examples had been built by the time the Armistice was declared in September 1943. Under its terms Piaggio and other Italian aircraft manufacturers were barred from production until the early 1950s. However, after this time aircraft were again built, the most notable types being the P136 twin-engined amphibian, the P148 and P149 single-engined trainers and the P166 twin-engined transport.

The Piaggio P32II medium bomber of 1938. It used Piaggio's own PX1 RC40 radial engines, each rated at 1,020hp.

The final wartime Piaggio aircraft design that actually flew was the P119 fighter; it had an eighteen-cylinder Piaggio radial engine buried in the fuselage, aft of the pilot.

Piaggio

Production of this new, small-wheel vehicle began at the end of 1945, at the giant Piaggio works at Pontedera, less than half an hour's drive from Pisa on the Gulf of Genoa. By the early 1950s the Vespa was not only a familiar sight on the roads of Italy, but also around the world. A number of licence agreements had also been signed – for example with Douglas in Britain, Hoffman of West Germany, and Moto Vespa SA of Spain.

There were also a number of World Speed records with a fully streamlined 18bhp, 123cc (42.9 × 44mm) machine.

1956 – THE MILLIONTH VESPA

Then, on 28 April 1956, came the millionth Vespa to roll off the production lines, which by then were turning out 10,000 scooters a month and employing a workforce of 4,000.

Throughout its first twenty years of Vespa production, Piaggio fought a fierce sales war with its arch-rivals Innocenti, who built the Lambretta.

The Vespa (Wasp) scooter series has sold in larger numbers than any other – and has been in continuous production for over fifty years.

Although Piaggio took over Gilera on 29 November 1969, it was not until two years later, in November 1971, that the first new models appeared. These ranged from 50 through to 175cc. But the smaller bikes were the mainstay. These were powered by an entirely new 49.77cc (38.4 × 43mm) single-cylinder two-stroke, with either four or five speeds.

Douglas – the Bristol Connection

By the early 1950s, the Vespa was not only a common sight in its native country but also all around Europe. Some of this was due to exports, but also because a number of manufacturing licences had been acquired; notably by Hoffman in Germany and Douglas in Great Britain. The latter, based in the Fishponds area of Bristol, also produced its own range of powered two-wheelers in the shape of the Douglas flat-twin series motorcycles, such as the MkV, 80 Plus, 90 Plus and finally the Dragonfly. All these machines were powered by a 348cc ohv engine with shaft final-drive.

The history of the Douglas marque had begun in 1906, when 27-year-old William Douglas purchased the design of a flat-twin engine from the Fairey company. Whilst Fairey ceased motorcycle production, Douglas went on to become one of the top British marques.

A Douglas engine was used by Freddie Dixon to gain an epic victory in the first ever Sidecar TT in the Isle of Man during 1923.

Post-war motorcycle production resumed in 1947 with the 348cc MkIII. Subsequently, this was replaced by the MkIV in 1949, followed by the MkV for the 1950 season.

Quite simply, the purpose of the Vespa licensing agreement was to bolster sinking demand for the motorcycle – a ploy which proved to be extremely successful, Douglas becoming the major Vespa manufacturer outside Italy by the mid-1950s, and even producing its own specialized variants of the Vespa theme.

With the acquisition of Gilera in 1969, Douglas was offered the UK distribution rights for the range of Gilera motorcycles. The main models imported during the early 1970s were the various 50cc and 125cc two-strokes and the 125/150cc Arcore ohv singles.

By the end of the 1970s Douglas (Sales and Service) Ltd, to give it its official trading title, was finding the going tough and, with the advent of the new decade, things got even tougher. Owners Westinghouse decided to cut their losses and get out of the two-wheel game. The result was that the London-based Heron company took over the Piaggio contract. But, unlike Douglas, they never assembled any machines, preferring to import the complete units from Italy (only Vespas were assembled by Douglas, not Gileras). Additionally, Heron were at first only interested in Vespa – they added Gilera later.

Douglas itself closed its doors at the beginning of 1982 for the final time.

During the 1960s the aviation side of the Piaggio group, which had been rebuilt successfully in the post-war era, was sold off. Then, on 29 November 1969, Gilera was passed from the receiver to Piaggio and a new era was born.

Although the Italian bike building business was at that time on its knees, Piaggio and its Vespa marque were flying high. Already established in scooters, and by now mopeds too, Piaggio was looking to the future and an involvement with motorcycles. A more healthy market was forecast, and Piaggio realized that to start from scratch would be far more expensive than purchasing an existing bike builder. Added to this, the Gilera name was one of the most respected in the industry.

VIANSON TAKES CHARGE

So Gilera, now under Piaggio control, and with the newly appointed Managing Director Enrico Vianson at the helm, strode forth into a new decade and made a new start.

A tentative range was assembled for 1970, sadly without the new 350/500 twins and 750 triple; more than a few

Piaggio

enthusiasts mourning the halt in their development. However, the Gilera company was more important, and with Piaggio's expertise and finance, plans were made. The years 1970 and 1971 were very much taken up with development of new models and building up the last of the 125/150/200 ohv singles. The only other machine produced and sold during this period was the Regolarita 125 ISDT bike.

THE 1971 MILAN SHOW

During the Milan Show held in November 1971, Gilera displayed a number of new models ranging from 50 through to 175cc. This translated into an eleven-model line-up for 1972. The new '50s' were all powered by an entirely new 49.77cc (38.4 × 43mm), with either four- or five-speeds. The alloy cylinder was inclined 15 degrees from the vertical with a radial finned alloy head. Ignition was provided by a crankshaft-mounted 6V, 18W flywheel magneto. Carburation was taken care of by a Dell'Orto SHA14-9 instrument.

Eventually Super, Touring, RS Trail and Enduro variants were offered. All shared a neatly crafted steel duplex frame with Ceriani exposed stanchion front forks and exposed spring rear shock absorbers. Depending upon the market, most models could be supplied with kick-starters or pedals and in varying states of engine tune. Only the Touring and Trail versions were officially imported into Britain. Power output figures ranged from 1.4 to 7bhp.

Piaggio's involvement with Gilera meant that the British Vespa importer also became the Gilera importer. Of the many models imported during the early-mid 1970s, perhaps the most popular was the 50 Touring moped with four-speeds. This, like the other sports mopeds, sold well until the British government changed the law, restricting sixteen-year-olds to 30mph 'slowpeds'.

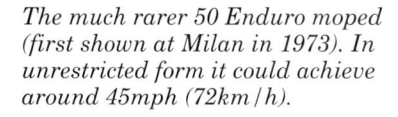

The much rarer 50 Enduro moped (first shown at Milan in 1973). In unrestricted form it could achieve around 45mph (72km/h).

Piaggio

Kickstart version of Gilera's 50cc Trail moped turned the machine into an ultra-lightweight motorcycle.

5v Touring (1972–1976)

Engine:	Air-cooled, single-cylinder two-stroke with piston-port induction Radial fin cylinder head
Bore:	38.4mm
Stroke:	43mm
Displacement:	49.77cc
Compression ratio:	5.5:1
Maximum power (at crank):	1.4bhp @ 4,700rpm
Lubrication:	Petroil
Ignition:	Flywheel magneto 6v
Fuel system:	Dell'Orto SHA14-9
Primary drive:	Gears
Final drive:	Chain
Gearbox:	Five-speed
Frame:	Duplex, steel cradle
Front suspension:	Telescopic fork, exposed stanchion
Rear suspension:	Twin shock, swinging arm
Front brake:	Full-width, drum
Rear brake:	Full-width, drum
Wheels:	19in front and rear
Tyres:	2.25 × 19 front and rear
Wheelbase:	46.25in (1,175mm)
Dry weight:	104kg (229lb)
Fuel tank capacity:	1.7imp. gal (7.7ltr)
Top speed:	25mph (40km/h)

Note:
There were several versions of the basic machine including the trial, touring (7bhp), RS and enduro. They were also offered with either kick-start or pedal, depending upon the country.

Piaggio

GILERA
50 c.c. MOPED 4 SPEED
(TOURING TYPE)

TECHNICAL DESCRIPTION

ENGINE: single cylinder 2 stroke, bore 38·4 mm stroke 43 mm capacity 49·797 cc compression ratio=12·1 max. power 4·2 DIN HP at 5,500 RPM. Lubrication=crank gear by petrol – oil mixture – gearbox and clutch by oil thrower, carburettor=float type Dell 'Orto SHB 18. Air filter paper element. Spark plugs=Bosch W 240 T2. Electric system=AC flywheel magneto. TRANSMISSION: Multi-plate clutch in oil bath. Constant mesh gearbox. Left-hand side foot gear shift lever 4 speed gearbox. Primary drive with helical gears 1/4,235 (17/72) secondary roller chain drive with rubber flexible coupling between crown and wheel. FRAME AND SUSPENSION: Double cradle tube frame. Front suspension. Double acting "Ceriani" hydraulic fork – road type. Rear suspension with swinging arm and shock absorbers. Front and rear light alloy brake drums with expanding brake shoes. Fuel tank capacity=1·5 galls. (approx.) tyres size=2¼" x 18". OVERALL DIMENSIONS AND WEIGHTS: Wheel base=1160 mm max. Length=1780 mm max. width 640 mm max. height 1040 mm min. ground clearance 180 mm steering angle from lock to lock position=90°. Dry weight= 68 Kg. PERFORMANCES: max. speed=65 Km/h (40 m.p.h. approx.) fuel=4% oil 30 (SAE) and 84 octane petrol mixture.

NOTE: The Company reserves the right to introduce any modifications to this specification without prior notice.

Ciao mopeds, Vespa motorscooters, Gilera motorcycles manufactured by PIAGGIO & C. Genoa, Italy.

Sole U.K. Concessionaires Douglas (Sales & Service) Limited, 2 Oak Lane, Fishponds, Bristol BS5 7XB
Tel: Bristol (0272) 654197 & (0272) 654882

Leaflet for the Touring model produced by the British Gilera importers, Douglas (Sales & Service) of Fishponds, Bristol.

There was also the competition-only 6V Competizione. Manufactured in both 49 and 74cc, these were sold in 1973 and 1974. The larger model – 74.56cc (47 × 43mm) pumped out an impressive 11.2bhp at 9,200rpm. Both versions had six-speed boxes and VHB Dell'Orto carbs (22mm on the 49, and 25mm on the 74). Except for engine size and carbs the two machines were identical.

ARCORE OHV SINGLE

Also making its début in 1972 was the Arcore, a four-stroke single with five-speeds and built in two engine sizes: 125 and 150. The bore and stroke measurements were the same as the models from the 1960s for their respective capacities. Built over a seven-year period, specification was similar on both models, with only details such as carb sizes (22 and 23mm VHB Dell'Ortos, respectively) separating them. Although the engines were based on the older singles there were notable differences, both internal and external.

Both the Arcore 125 and 150 were imported into Britain via Douglas Sales and Service of Fishponds, Bristol, one of their salesmen being no less than 'Titch' Allan, the founder of the VMCC (Vintage Motor Cycle Club).

Although the backbone of the Gilera range in the 1970s was largely made up of the 50 two-strokes in their various guises and the Arcore four-stroke singles, the company did bring out a series of other models, some of which were destined for only a brief life. These included the 1975 CB1 (a commuter moped with single seat and four-speeds), the 1977 CBA (basically the same machine but with automatic transmission), the 1978 TS50 (ultra-lightweight motorcycle with five-speeds) and the GR2 and Trial 50s (both trial machines with six-speeds).

A NEW 125 SERIES ARRIVES

Of more significance was a new 125 class engine which was used to power a variety of motorcycles and made use of the technology used by Gilera in their off-road, but not directly based on the actual, trials/moto cross unit. The 122.48cc (57 × 48mm) featured an alloy head, Ducati electronic ignition (35W), Dell'Orto 24BS carb, five-speeds and geared primary drive. The first two machines, both launched in 1977, were the GR1 (Trail, 16.4bhp at 7,300rpm) and the TG1 (Roadster, 14.5bhp at 7,300rpm).

THE T4S 200

Also new for 1977 was the T4S 200. This model used an ohv engine based on the Arcore series, but with a larger capacity of 198.42cc, with both the bore and stroke

British journalist, the late Bob Currie, looks happy as he sits astride a 150 Arcore flanked by members of the Piaggio management, circa 1976. The Arcore, in 125 and 150cc guise, ran from 1972 through to 1979.

Piaggio

	5v Arcore 125/150 (1972–1979)
Engine:	Air-cooled, single-cylinder, four-stroke, ohv with parallel valves
Bore:	60mm
Stroke:	125 44mm; 150 54mm
Displacement:	125 124.4cc; 150 152.68cc
Compression ratio:	10:1
Maximum power (at crank):	125 12bhp @ 8,500rpm; 150 14bhp @ 8,300rpm
Lubrication:	Wet sump
Ignition:	Flywheel magneto 6v
Fuel system:	125 Dell'Orto VHB22; 150 Dell'Orto VHB23
Primary drive:	Gears
Final drive:	Chain
Gearbox:	Five-speed
Frame:	Duplex, steel cradle
Front suspension:	Telescopic fork, with rubber gaiters
Rear suspension:	Twin rear shock, swinging arm
Front brake:	Drum, full-width, aluminium, SLS
Rear brake:	Drum, full-width, aluminium, SLS
Wheels:	18in front and rear
Tyres:	Front 2.75 × 18; rear 3.00 × 18
Wheelbase:	50in (1,280mm)
Dry weight:	125 240lb (109kg); 150 258lb (117kg)
Fuel tank capacity:	2.5imp. gal (11ltr)
Top speed:	125 69.5mph (112km/h); 150 74.5mph (120km/h)

(Above) *The 125 shown also has indicators, not found on all Arcore models. Maximum speed on the smaller model was 69mph (111km/h), the 150 being capable of 75mph (121km/h). Both had five-speed gearboxes.*

(Below) *Side-view drawing of the Arcore engine unit.*

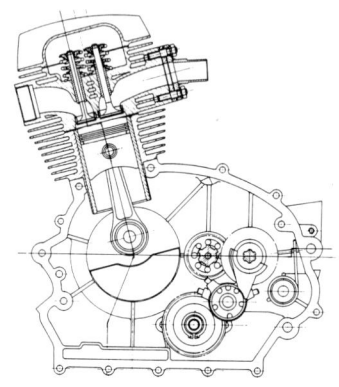

Piaggio

enlarged (66 × 58mm). Running on a 10.9:1 compression ratio, the 200 offered its rider 17bhp at 8,000rpm. The styling was similar but not identical to the TG1, and the larger bike came with cast alloy instead of wire wheels; a custom version arrived later, being shown at the Milan Show in late 1981.

NEW MODELS ARRIVE

In 1982 came an updated TG1 with oil pump lubrication and, later still, the TG2 and TG3. These all used updated versions of the TG1 engine in the same basic chassis, but improvements to areas such as braking, better wheels, improved switch gear and more modern styling. The TG3 was a custom model with similar styling to the four-stroke T4 Custom.

Other models to appear in the early 1980s included the Eco (automatic two-stroke moped with 12in cast alloy wheels), the Vale (ditto, but with 16in wheels) and the GSA (scooter with the same horizontal 49.82cc, 38.4 × 43mm automatic engine as the mopeds).

1984

Then in 1984 came the landmark RX/RW LC (Liquid Cooling) trail bikes with enduro styling, soon to be named Arizona. These, together with the other liquid-cooled Gilera two-strokes of the modern era, are described in Chapter 8. But it had been the various air-cooled models in both two and four-stroke guises which had carried Gilera back into profit during the immediate post-Piaggio take-over years.

(Left) *In 1977 Gilera brought out an entirely new two-stroke to augment its sales in the all-important 125cc categories; this was the TG1 (Turismo Gilera One). Like all too many Italian lightweight two-strokes of the era, the TG1, powered by a 122.48cc (57 × 48mm) five-speed engine, suffered from feeble 6v electrics, the messy petroil mixture and substandard paintwork and chrome. Additionally, in overseas markets such as Britain, it usually had a higher price tag than the Japanese opposition. The early TG1 had wire wheels; after 1980 these were changed to cast alloy (as shown).*

(Right) *At the same time as the TG1 was hitting the streets, Gilera also launched an on-off road version, the GR1 (Gilera Regolarita One). The GR1 was styled to attract youngsters, but in the engine department it was identical to the TG1; however, although it shared the same basic frame, the majority of its running gear was different. There were Marzocchi gas rear shocks, injection moulded plastic guards, hi-level black (easy rust!) exhaust, 140mm Grimeca moto cross-type front brake, braced bars, rubber mounted rear light, lower gearing and knobbly tyres with high tensile chrome-plated wheel rims.*

Piaggio

(Left) *The 50TS (Turismo Strada) sports moped was essentially a development of the earlier RS model, itself stemming from the original Touring and Super 50's which had made their début back in 1972. The first 50TS appeared in 1978 with wire wheels and a somewhat spartan specification, but this higher specification model was imported into Britain by Douglas of Bristol and had the following 'extras' as standard: disc front brake, cast alloy wheels, dual seat, direction indicators and pillion footrests.*

(Below) *The largest four-stroke offered by Gilera during the 1970s (except for the old-stock 202 from the 1960s) was the T4S, seen here at the German Cologne Show in September 1978. Clearly developed from the Arcore series, the 198.42cc (66 × 58mm) engine was also used for a custom version that Gilera displayed at the Milan Show in November 1981. Neither sold in any real numbers.*

(Above) *The 50TS 49.8cc (38.4 × 43mm) engine had piston port induction and a five-speed gearbox. The alloy cylinder was flash-chrome plated, there was a radially-finned cylinder head and rubber blocks to cut down 'fin-ringing' on both the cylinder and head. Power output was a miserly 1.48bhp at 4,500rpm.*

7 Dirt Bikes

Gilera has always had a proud history, not just of building race winning Grand Prix bikes and robust sporting roadsters, but also for its long line of successful dirt bikes.

EARLY EFFORTS

Their first taste for off-road competition came via entry into the famed *Sei Giorni* (Six Days). First staged in 1913, by the mid-1920s Gilera was making an impact thanks to men such as Gino Zanchetta, Rosolino Grana, Miro Maffeis, Umberto Meani and Luigi Gilera, the last four making up the official works team for several years.

Machinery used was invariably the 500 single in both ohv and sv form. Besides the solos, Gilera also entered sidecar outfits, Luigi Gilera's own speciality.

Even though its off-road efforts were overshadowed by the fire-breathing, four-cylinder road racers, Gilera continued its ISDT efforts into the post-war era. In addition, the Saturnos made a significant impact in the field of motocross racing (*see* Chapter 2).

HEAVYWEIGHT SINGLES

By the mid-1950s the factory's efforts were being channelled away from its large heavyweight singles and into its new

Gilera could trace a long distinguished history in the ISDT (International Six Days Trial). For example, during the 1920s men such as Zanchetta, Grana, Maffeis, Meani and Giuseppe Gilera's brother Luigi, all competed with great success.

Dirt Bikes

Post-war this trend continued as Mario Masserini (247cc Gilera Nettuno) shows during the 1948 ISDT in which he won a gold medal.

range of small capacity, ohv parallel valve singles with unit construction engines. These first appeared in action – in 175 form – at the 1956 ISDT, held at the German ski resort of Garmisch-Partenkirchen. The new Gileras helped Italy to score the runner-up position in the all-important Trophy contest. Two of the five-man Italian squad were Gilera-mounted, D. Fenocchio and P. Carissoni.

GOLD MEDAL SUCCESS

Gilera also entered a team in the Manufacturers' awards (Gian Franco Saini, plus Fenocchio and Carissoni). All three riders also gained gold medals.

By 1961 the *Sei Giorni* bikes also included 98 and 124 models. All were clearly based on existing standard production roadsters but with many special features, enabling them to be reasonably successful.

But it was not until the Piaggio takeover at the end of the decade that Gilera was to make a real impact in the off-road sector.

During the immediate post-war period it was not uncommon for the factory's Grand Prix stars to also take part in the 'Six Days'. Here Carlo Bandirola stops for a drink during the final day of the 1951 event at San Remo.

Dirt Bikes

At the German ski stadium in Garmisch-Partenkirchen riders prepare their machines for the 1956 ISDT. In the foreground is Pietro Carissioni's 175 Gilera. This marked a serious return by the Arcore factory into the event. The renewed Gilera effort helped Italy to runner-up position in the all-important Trophy contest that year.

One of the ex-works 1956 ISDT bikes on display at an Italian classic bike show during the mid-1980s.

125

Dirt Bikes

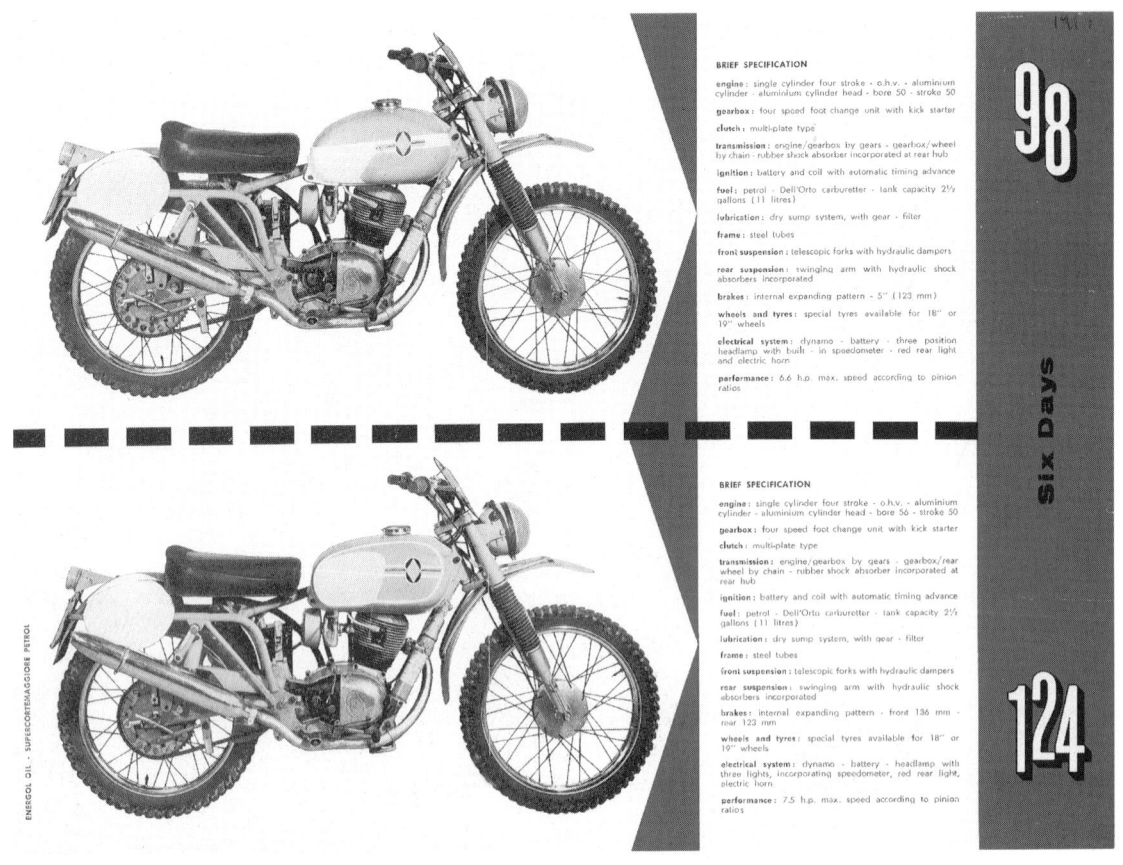

Brochure issued by Gilera illustrating their customer versions of the successful ISDT bikes, in 98 and 124 variations. Both were medal winners in the right hands and helped to keep the factory's sporting reputation alive following its withdrawal from Grand Prix road racing.

THE REGOLARITA

Even before this, in 1968, the Arcore factory had built a new 125-class off-roader. Called the Regolarita, it had a rotary valve two-stroke engine; the square 54 × 54mm dimensions giving 123.670cc. This first stab at a modern two-stroke provided 20bhp at 9,500rpm. The gearbox was a five-speeder with primary drive by gear; a 100cc version was also built. Both featured in the Italian Enduro Championships of 1969 and 1970, the larger model piloted by Fausto Vergani.

This same rider was victorious in the 1970 Valli Bergamassche trial for Gilera.

In 1973 an all-new Regolarita 125 made its début. About the only thing it shared with its predecessor was its 54 × 54mm bore and stroke dimensions and its use of rotary valve induction – however, the latter was now totally different in its location and design. Although maximum power of 20.8bhp at 9,200rpm was little different on paper, the motor was far more flexible and stronger. Carb size had been upped from 26 to 30mm and an additional ratio added to its gearbox. At 207lb (94kg) dry weight it was some 11lb (5kg) lighter than the earlier bike. This was largely achieved by the use of magnesium outer engine covers and the increased use of plastics for components such as mudguards and side panels.

1973 EUROPEAN ENDURO CHAMPIONS

Gilera's managing director Enrico Vianson quickly saw the potential for success, and thus publicity, for the new enduro model and duly signed some of the best Italian riders of the day. This faith was immediately repaid by Gilera's victory in the 1973 European Championship two-day enduro series.

ISDT was fast approaching and the factory was looking forward to equipping the Italian national team in the Trophy section, however, a disappointment was in store. The Italian Federation instead decided to field a team of riders on Austrian KTM models; presumably preferring to choose a proven design, rather than something as new as the latest Gileras. So the Arcore marque chose to concentrate on the Silver Vase. Some unfortunate incidents transpired to prevent ultimate success, but Gileras did make an impres-

Following Piaggio's take-over in November 1969, the new management re-launched its interest in long-distance trials. Here, members of the 1973 Italian Silver Vase team pose prior to the ISDT that year. All were mounted on the brand new Regolarita Casa model. Powered by a specially developed 123.670cc (54 × 54mm) single-cylinder rotary valve engine with six-speed transmission, these impressive newcomers produced 20.8bhp at 9,200rpm, giving a road speed in excess of 70mph (112.5km/h). Although not entirely successful in the ISDT, the new Gilera dominated much of the Italian national scene that year, with Alessandro Gritti taking the Senior Enduro title.

One of the works Regolarita Casa enduro bikes. For 1974 a 175 version was also raced. This had new 61 × 59.5mm bore and stroke dimensions, giving 173.84cc. Like its smaller brother it had rotary valve induction and six speeds. The 125 (shown here) weighed in at 207lb (94kg), the 175 216lb (98kg). Then, after building such a competitive pair of dirt irons, Gilera shocked everyone by scrapping its factory enduro team.

Following Gilera's withdrawal at the end of 1974, the Elmeca company built a number of Gilera-inspired designs under licence. Typical of these is this 1975 Elmeca-Gilera 125 Regolarita. Developed from the famous ISDT machines it was a leading contender for honours until outpaced by a new breed of enduro bike from the likes of SWM and KTM from 1977 onwards.

sive showing none the less. Although, in the end, they only won the 500cc class outright, on the fifth day they led the 50, 75 *and* 125cc categories! There was no doubting their future potential.

ELMECA

Much of the development was done in conjunction with the Cafasse-based (near Turin) Elmeca concern. They were established off-road specialists who had also been involved with Gilera before the Piaggio buy-out. In fact customer-versions of the trials bikes were sold under the Elmeca-Gilera brand name.

The 50 and 75 models were clearly based on the standard production 50 roadster and trail bikes. They were also not only supplied to factory-supported riders but sold through the Gilera dealer network. Both were known as the 6V Competizione (*see* Chapter 6). They were comprehensively equipped with every conceivable off-road goody including map-case, rubber-mounted rear light, quickly detachable wheels, Boranni alloy wheel rims, centre and side stands, braced handlebars, six-speed gearbox, hi-level exhaust, large capacity air filter, tank breather, competition number backgrounds, knobbly tyres (21in front, 18in rear) and a dry weight of only 198lb (90kg).

Dirt Bikes

Competizione 50/75 (1973–1974)	
Engine:	Air-cooled, single-cylinder, two-stroke, with piston-port induction. Radial fin cylinder head
Bore:	50 38.4mm; 75 43mm
Stroke:	43mm
Displacement:	50 49.77cc; 75 74.56cc
Compression ratio:	50 13.5:1; 75 14.5:1
Maximum power (at crank):	50cc 8.8bhp @ 10,750rpm; 75cc 11.2bhp @ 9,200rpm
Lubrication:	Petroil
Ignition:	Flywheel magneto, 6v 28w
Fuel system:	50 Dell'Orto VHB22; 75 Dell'Orto VBH25
Primary drive:	Gears
Final drive:	Chain
Gearbox:	Six-speed
Frame:	Duplex, steel cradle
Front suspension:	Ceriani telescopic fork, exposed stanchion
Rear suspension:	Twin shock, swinging arm
Front brake:	Full width hub, drum
Rear brake:	Quickly detachable, conical, drum
Wheels:	Front 21in; Rear 18in
Tyres:	Front 2.50 × 21; Rear 50cc 3.00 × 18; 75cc 3.50 × 18
Wheelbase:	50.8in (1,290mm)
Dry weight:	198lb (90kg)
Fuel tank capacity:	1.5imp. gal (7ltr)
Top speed:	50cc 53mph (86km/h); 54.5mph (88km/h)

A NEW 175 MODEL

Building on the achievements gained in 1973, competition activity continued in 1974, and a new, larger model was added to the team, a 175. Although based on the successful 125, the larger dirt iron had the different bore and stroke measurements of 61 × 59.5mm, giving 173.840cc. With a dry weight of only 216lb (98kg) (8.8lb (4kg) more than the 125), the '175' offered excellent power-to-weight ratio. Maximum power of 24bhp generated at 10,000rpm, making it a match for any machine in its class at the time. This capability was reflected by Gilera's ability to attract the very best riders.

The first two-day event in the 1974 European Championship calendar saw Gilera riders finish first in the 500cc, first in the 75, fifth in the 100 (still on 75s!), first in the 125 and third in the 175cc. Results like these continued throughout the season, and Gilera was nominated by the Italian Federation for both the Trophy and Vase squads in that year's ISDT.

DISAPPOINTMENT

Gilera was on course for overall victory in both the Trophy and the Vase, but in the end runner-up in the Vase contest was the best they could achieve. Their trophy dreams were doomed when, on the Thursday, Fausto Oldrati's rear wheel began breaking up. This gradually got worse, until he began losing marks, effectively letting the Czechs through to gain yet another Trophy victory, Italy eventually finishing third.

Regolarita 175 (1974)

Engine:	Air-cooled, single-cylinder, two-stroke, with reed valve induction via crankcase
Bore:	61mm
Stroke:	59.5mm
Displacement:	173.840cc
Compression ratio:	12.5:1
Maximum power (at crank):	24bhp @ 10,000rpm
Lubrication:	Petroil
Ignition:	Electronic
Fuel system:	Dell'Orto PHB34 carb
Primary drive:	Gear
Final drive:	Chain
Gearbox:	Six-speed
Frame:	Duplex, steel cradle
Front suspension:	Ceriani, exposed stanchion central axle forks
Rear suspension:	Twin shock, steel swinging arm
Front brake:	Conical, drum
Rear brake:	Conical, drum
Wheels:	Front 21in; Rear 18in
Tyres:	Front 3.0 × 21; Rear 4.0 × 18
Wheelbase:	54.5in (1,386mm)
Dry weight:	216lb (98kg)
Fuel tank capacity:	2.2imp. gal (10ltr)
Top speed:	79mph (127km/h)

Even so the results were good, taking into account the fact that Gilera had only recently returned to world-class competition. To take on the Germans and Czechs, who had dominated this branch of the sport for so many years, was no mean achievement.

THE BOMBSHELL

Then came the bombshell – Piaggio decided that Gilera should withdraw from all forms of motorcycle sport – including the ISDT.

1975 should have been a fantastic year for the factory; instead it concentrated its efforts on developing its standard production range and improving its dealer network.

Pure competition models continued, but only under the Elmeca name in both trials and motocross, the latter first appearing in 1976. The wording on its fuel tank reading 'ELMECA powered by Gilera'.

OVER-THE-COUNTER MODELS

For 1976 Elmeca sold both the Regolarita (Enduro) at 1,476,300 lire and the Cross (Motocross) at 1,539,000 lire. Both used a revised version of the 1974 ISDT engine. The stroke was slightly shorter at 53.5mm (formerly 54mm), the bore remaining at 54mm giving a capacity of 122.7cc. There was also improved power output, the enduro putting out 23.5bhp at 9,750rpm, the motocross 24bhp at 10,300rpm. Both machines used the same Dell'Orto PBH 32AS carb.

The Elmeca-Gilera 125 Cross was built from 1976 through to 1979. This used a tuned version of the Elmeca-Gilera enduro engine. It differed from the original works Gilera models in having a shorter stroke of 53.6mm, instead of the original 54mm. This gave an engine size of 122.7cc. The motocrosser produced its maximum power of 24bhp at 10,300rpm. It offered a viable Italian alternative to the swarms of Japanese dirt racers of the era. Dry weight was 216lb (98kg).

Both these bikes were sold (in updated form) until 1979, by then the prices had risen to 1,618,800 lire (Regalorita) and 1,653,000 lire (Cross)

The Elmeca-Gilera had won the 1978 Italian Motocross Championship with Dario Nani.

WITTEVEEN ARRIVES

Thanks to this success, Piaggio again allowed Gilera to take part officially. This in turn put things into a higher stratum; enter Jan Witteveen. This Dutch two-stroke specialist joined Gilera in 1980, taking over the responsibility of competition bike design for the Arcore factory. A new 125 with *water-cooling* soon appeared. Witteveen's expertise was applied to making the existing Gilera engine more efficient, rather than designing a completely new unit. Thus the 54 × 53.6mm bore and stroke dimensions from the Elmeca-Gilera of the late 1970s were retained.

WATER-COOLING

One of these water-cooled 125 'crossers was campaigned in the 1981 and 1982 Italian and World Championships by factory rider Michele Rinaldi – with some considerable success. The works models, as raced by Rinaldi, had Marzocchi single-shock rear suspension. Gilera also sold customer variations; the first of these, the Cl Competizione Cross was marketed in 1981 and 1982. This featured water-cooling of the engine, but came with different rear shock (Corte Cosso) and Marzocchi, leading axle 38mm forks. There was also the El Competizione enduro model.

Dirt Bikes

This is the final Elmeca-Gilera Cross as sold during 1978 and 1979. Besides its more modern styling, it also had extra power (23.5bhp) and less weight (202lb (92kg)) than the original.

A previously unpublished photograph of probably the most exotic motocross design ever, the experimental Gilera Bicilindrica *(twin cylinder) of 1981. The 124.88cc rotary valve engine had its tiny 43 × 43mm cylinders staggered one above each other, not V-shaped as many might consider. The work of Jan Witteveen, the twin was not put into production, even though with 36bhp at 12,000rpm it was more powerful than the single.*

El Enduro (1981)

Engine:	Liquid-cooled, single-cylinder, two-stroke, with reed valve induction
Bore:	54mm
Stroke:	53.6mm
Displacement:	122.75cc
Compression ratio:	15:1
Maximum power (at crank):	31.3bhp @ 10,000rpm
Lubrication:	Petroil
Ignition:	Electronic
Fuel system:	Dell'Orto PHBE36HS carb
Primary drive:	Gear
Final drive:	Chain
Gearbox:	Six-speed
Frame:	Duplex, steel cradle
Front suspension:	Marzocchi 35 telescopic fork, with leading axle
Rear suspension:	Twin shock, swinging arm
Front brake:	125mm drum
Rear brake:	125mm drum
Wheels:	Front 21in; Rear 18in
Tyres:	Front 3.0 × 21; Rear 4.0 × 18
Wheelbase:	56.9in (1,445mm)
Dry weight:	187lb (85kg)
Fuel tank capacity:	1.48imp. gal (6.75ltr)
Top speed:	89mph (140km/h)

For 1983, whilst Rinaldi carried Gilera's hopes at the highest level, the customer motocrosser became the C2 with revised styling and the single-shock rear suspension as pioneered on Rinaldi's bike. Gilera also built and tested a 125 *twin* motocrosser in 1981 ridden by Maurizio Perfina; the 124.88cc (43 × 43mm) produced 36bhp at 12,000rpm.

Gilera themselves made a return to dirt-bike sport in 1980, by way of a new water-cooled, 125-class motocrosser, designed by the Dutch engineer Jan Witteveen. It used the same engine dimensions as the Elmeca-Gileras, but was otherwise almost entirely new. Running on a 15:1 compression ratio it produced a class-challenging 31.3bhp at 10,000rpm. From 1981 Michele Rinaldi rode for Gilera in the World Championship with some considerable success.

Dirt Bikes

Bicilindrica (Twin Cylinder) 125 Motocross (1981)

Engine:	Liquid-cooled, twin-cylinder, with stacked cylinders and reed valve induction
Bore:	43mm
Stroke:	43mm
Displacement:	124.88cc
Compression ratio:	16:1
Maximum power (at crank):	36bhp @ 12,000rpm
Lubrication:	Petroil
Ignition:	Electronic
Fuel system:	Twin 28mm Dell'Orto carbs
Primary drive:	Gear
Final drive:	Chain
Gearbox:	Six-speed
Frame:	Duplex, steel cradle
Front suspension:	Marzocchi telescopic fork, leading axle type
Rear suspension:	Monoshock, with steel swinging arm
Front brake:	Drum 125mm
Rear brake:	Drum 125mm
Wheels:	Front 21in; Rear 18in
Tyres:	Front 3.0 × 21; Rear 4.0 × 18
Wheelbase:	58in (1,480mm)
Dry weight:	207lb (94kg)
Fuel tank capacity:	1.47imp. gal (6.7ltr)
Top speed:	93mph (150km/h)

On the enduro front Gilera had built an updated machine at the beginning of 1982. Called the Series K, this now sported Sachs Hydrocross rear suspension and specially commissioned Gilera forks based on the production Marzocchis. In this form Gilera dominated much of the domestic Italian enduro scene during 1982 and 1983.

A production version of the water-cooled Gilera motocross was built in a variety of models starting with the CI Competizione in 1981. Unlike the factory model raced by Rinaldi, the customer versions had less power, but increased reliability ...

Dirt Bikes

1983 MILAN SHOW

At the Milan Show in November 1983 the company displayed its latest production off-road bikes – the motocross HX 125LC (now producing a claimed 32bhp at 10,700rpm) and the HE125LC enduro, the latter for the first time sporting single-shock link-type rear suspension.

250 MOTOCROSS

The next step was upwards, into the 250 motocross category. Clearly based on the HX125LC, the HX250LC (248.939cc, 71.5 × 62mm) shared many of the cycle parts from the smaller model, including a disc front brake.

Running on a 15:1 compression ratio, the rotary valve engine pumped out an impressive 42bhp at 7,750rpm. Strangely, both the 125 and 250 HX model used the same carburettor, a 36mm Dell'Orto. Both also used a Nikasil-coated cylinder barrel (introduced in 1981 by Witteveen on the original factory water-cooled 125 motocrosser).

... this is the CI. Technical details included a 122.7cc (54 × 53.6mm) disc valve engine producing 31.3bhp through a six-speed gearbox.

In parallel with its motocross efforts, Gilera also built and sold an enduro version, the EI. Again, first offered in 1981 (shown), it was later produced from 1982 as the K-series with a Sachs Hydrocross rear shock in place of the original Corte Cosso unit.

Dirt Bikes

HX250LC Motocross (1984)

Engine:	Liquid-cooled, single-cylinder, two-stroke with reed valve induction
Bore:	71.5mm
Stroke:	43mm
Displacement:	124.88cc
Compression ratio:	16:1
Maximum power (at crank):	36bhp @ 12,000rpm
Lubrication:	Petroil
Ignition:	Electronic
Fuel system:	Twin 28mm Dell'Orto carbs
Primary drive:	Gear
Final drive:	Chain
Gearbox:	Six-speed
Frame:	Duplex, steel cradle
Front suspension:	Marzocchi telescopic fork, leading axle type
Rear suspension:	Monoshock, with steel swinging arm
Front brake:	Drum 125mm
Rear brake:	Drum 125mm
Wheels:	Front 21in; Rear 18in
Tyres:	Front 3.0 × 21; Rear 4.0 × 18
Wheelbase:	58in (1,480mm)
Dry weight:	207lb (94kg)
Fuel tank capacity:	1.47imp. gal (6.7ltr)
Top speed:	93mph (150km/h)

In 1984 Gilera launched the HX250LC motocrosser. Powered by a 248.939cc (71.5 × 62mm) rotary valve engine it gave 42bhp at 7,750rpm. Then, at the Milan Show in November 1985, an improved version, the NX250LC appeared. Power was now up to 44bhp with a host of other updates.

Works version of the production RX250 Rally enduro, which had been introduced at the end of 1984, on the company's stand at Milan in 1985. During 1986 it proved Italy's best selling 250 enduro bike with no less than 349 examples being snapped up by enthusiastic customers.

PARIS–DAKAR

By 1985 Gilera had also built a Paris–Dakar Rally-type enduro mount. One of these was displayed at the Milan Show that year covered in mud gained during one such event. The production version was marketed (as the RX250 Rally) in 1986, when it proved to be the top seller (349 examples) in Italy that year in the 250 enduro category. The 1986 'crosser was now called the NX250LC and offered 44bhp.

All this hard work in the competition area was also to prove of great benefit to Gilera who were able to bring out a new family of two-stroke sports roadster and trail models, using the technology gleaned from the white-hot cauldron of dirt bike competition during the early and mid-1980s (*see* Chapter 8).

8 Two-Strokes – the New Breed

What could be classed as Gilera's 'new breed' of non-competition production two-strokes to be offered to the public was the RX125 trail bike of 1984 (first shown in prototype form at the 1983 Milan Show). However, although it benefitted from the company's off-road sporting involvement in that it utilized the technology, liquid cooling and single-cylinder used by its sporting brothers, its detailed specification was not duplicated.

RX125 SERIES

At the heart of the RX125LC (also sold in higher spec. RW125LC form) was its 124.3cc (65 × 50.5mm) reed induction, six-speed engine. This produced 19bhp at 7,750rpm (RW125LC, 20bhp at 8,000rpm). Both versions shared a 26mm Dell'Orto carb. and a wheelbase of 1,385mm.

Also of interest was the disc (front) and drum (rear) brakes, 21in (front) and 16in

The very first of Gilera's new breed of standard production two-stroke models was the RX125 trail bike (shown at the Milan Show in late 1983 and on sale the following year). Soon the RW125 Arizona appeared in higher spec form. By 1985 this was known as the Arizona Hawk. All featured a liquid-cooled 124.3cc (65 × 50.5mm) engine with reed valve induction and six speeds.

Two-Strokes – the New Breed

(rear) wheels, Gilera 'Monodrive' single shock rear suspension – unusually this had the linkages for this *outside* the swinging arm, with the vertically positioned shock mounted centrally.

In fact the whole machine looked very similar to a Suzuki TS125 of the same era. The RW125 had an 'Arizona' badge on either side of the tank, rather than the Gilera logo as on the RX version.

1985

1985 was a turning year in Gilera's fortunes, with a massive development programme, resulting in a host of new production models. The Milan Show that year emphasized that Gilera was back as a major force with one of the biggest and brightest stands at the exhibition. The two-stroke line-up comprised the following models: 125 Arizona Hawk (trail), 125RTX (trail), 125KZ (super sports), 125KK (super sports), 125RV (sports), 200 Arizona Hawk (trail), 200RTX (trail), 200RV (sports), 250NGR (sports) and 250 Arizona Rally (enduro). The vast majority of these machines were equipped with an electric starter as standard equipment.

A pure street bike soon appeared as the 125RV, followed by the 200RV and also the NGR250 (shown). This latter bike had its 248.94cc (71.6 × 62mm) power unit directly derived from the HX250LC motocrosser. The remainder of its specifications included balancer shaft, triple disc brakes, integral fairing, electric start, anti-dive Marzocchi forks, 16/18in front/rear wheels and a maximum speed of 106mph (170.5km/h).

The 1985 125RV – compare the differences with its larger brother above.

Two-Strokes – the New Breed

250NGR (1984–1986)	
Engine:	Liquid-cooled, single-cylinder, two-stroke with valve induction
Bore:	71.5mm
Stroke:	62mm
Displacement:	248.939cc
Compression ratio:	13.5:1
Maximum power (at crank):	38bhp @ 7,750rpm
Lubrication:	Oil pump and separate oil tank
Ignition:	Electronic
Fuel system:	Dell'Orto 32mm carburettor
Primary drive:	Gears
Final drive:	Chain
Gearbox:	Five-speed
Frame:	Steel, square tube, duplex.
Front suspension:	Telescopic fork, Marzocchi 38mm diameter stanchion
Rear suspension:	Square section steel swinging arm with single shock; 140mm travel
Front brake:	Twin 240mm discs with Brembo two-piston calipers
Rear brake:	Single 240mm disc with Brembo two-piston caliper
Wheels:	Front 16in; Rear 18in
Tyres:	Front 100/90 × 16; Rear 110/90 × 18
Wheelbase:	53in (1,350mm)
Dry weight:	304lb (138kg)
Fuel tank capacity:	4.4imp. gal (20ltr)
Top speed:	100mph (160km/h)

The 125 trail mounts were developments of the RX/RW series, but with more power (22bhp at 8,800rpm). The 26mm carb size was retained, but now featured a power valve to improve mid-range ability.

KZ AND KK SPORTS MODELS

The new sports roadster 125s (KZ and KK) produced yet more power (26bhp at 9,000rpm). The main difference between the two models was the fairing (a top section and belly pan on the KZ and a full fairing on the KK). There was also a bigger fuel capacity on the KK – 3.3imp. gal instead of 2.64 (15ltr instead of 12). Both models shared the following features: 13.5:1 compression ratio; gear ratios: 1st 2.7692, 2nd 1.866, 3rd 1.4-4, 4th 1.13-5, 5th 0.96-6, 6th 0.883; Forcella Italia (formerly Ceriani) 36mm front forks whilst the Gilera 'Power-Drive' rear suspension featured a single shock and linkage, high-level exhaust system, 16in cast alloy wheels, 240mm single disc brakes front and rear, and an aluminium square-tube mainframe.

NATIONAL RACING SERIES

Gilera also organized a national racing championship series called the Gilera 125KZ Endurance. This one-model series proved very popular and was a forerunner to the highly successful SP (Sport Production) series that bred the next generation

Two-Strokes – the New Breed

(Above) *Gilera's best selling model of the mid-1980s was the KZ-KK series. These 125 sportsters took Italy by storm and during 1986 no fewer than 16,000 KZs alone were sold in Italy that year.*

(Left) *The KZ-KK 124.3cc (65 × 50.5mm) reed induction engine, with liquid cooling.*

Two-Strokes – the New Breed

of World Championship contenders – Max Bioggi and Loris Capirossi to name but two.

For example, the 1988 series was run over seven rounds, these being: Monza 17 April, Magione 1 May, Varano 15 May, Mugello 26 June, Misano 15 August, Vallelunga 18 September and Mugelo 25 September.

MORE NEW MODELS

The 125RV and 200RV were essentially the same basic bike, except for the engine assembly. But even here they were closely related, the '200' was actually a 183.4cc achieved simply by increasing the 125's cylinder bore from 56 to 68mm – the 50.5mm stroke remaining unaltered. Even the carburettor size was the same (26mm) on both models. The 200RTX and 200 Arizona Hawk both used the same engine as the 200RV roadster, but in 125 trailsters' chassis; Gilera thus making full use of its parts stock.

The 250NGR and 250 Arizona Rally again followed this trend, but in their case the same engine and carb was utilized for both models. This being a 248.939cc (71.6 × 62mm) directly derived from the HX250LC motocrosser. Both the NGR and Arizona Rally ran on a compression ratio of 13.5:1 and employed a 32mm Dell'Orto carb (the motocrosser used a 36mm instrument).

BALANCE SHAFT

Both also used a five-speed box (as did the pure dirt bike) and a balance shaft was filled to smooth out otherwise annoying vibes. The NGR borrowed its styling (and even its red/black colour scheme) from the then current Kawasaki GPZ series. But not their power output ... the Gillie only producing 38bhp at 7,750rpm (the Arizona Rally producing the same power, but at the slightly higher 8,000rpm).

During 1986 the 125KZ emerged as Gilera's best seller, some 16,000 being sold in Italy during that year. But strangely, when the Heron Corporation began importing the bikes that year (marketing

There was also a very successful one-make race series, the KZ Endurance. This was to act as the forerunner to the even more successful SP (Sport production) national series that has since bred a new generation of Italian World Champions.

them through London-based Derek Loan Motorcycles – a.k.a. Gilera London) it chose to ignore this bike, instead bringing in the RV125, 250NGR and RX125 Arizona.

Performance Bikes tested the latter bike in their November 1986 issue. A selection of their comments follows:

> Compared with its predecessor, [the same magazine had tested an air-cooled GR1 back in 1979] the 1986 Gilera trail bike is in another league; like the switchgear and electrics, the clocks were almost up to Japanese standards. If you compare the Arizona with any of its Japanese rivals you'll soon realize it's not giving away anything in terms of price or specifications. (How many of the opposition have an electric start?) The finish is as good as the Orientals too, with lots of gleaming allen screws and nyloc in all the right places. And if you want to pose, exclusivity beats almost anything.

In fact the only things *Performance Bikes* didn't go a bundle on were an over-noisy clutch, 12bhp which really was 12bhp (for the British learner laws) and the fact the Gilera didn't put their name on the tank – only the Arizona logo.

At £1,199 the Arizona compared well with the Oriental opposition (TS125X £1,219, DT 125LC £1,158, KMX125 £1,149 and MTX125 £1,159). At 64mpg (4.4l/100km) (test average) the Italian was also the most economical too.

The Bologna Show in December 1986 was the launch pad for yet more new two-strokes.

Motor Cycle News described the new Gilera Fast Bike as a 'Style Mixer'. In fact there were two Fast Bikes – a 25bhp 125, and a similarly powerful but more torquey 200. The idea was to offer a custom bike for the road rider who went for the off-road look.

The water-cooled, single-cylinder, two-stroke engines were mounted in box section mono-shock frames with disc brakes front and rear, and leading axle front forks.

By the time the 1987 Milan Show came around there were no less than an amazing *twenty-four* production Gilera two-strokes (including motocross and enduro models). The full line-up was as follows: Mopeds: CBA, CBI, ECO, RI; Scooters: GSA; 125 Motorcycles: RV, Hawk, RTX, KZ, KK, ER, RRT, Fast Bike, Rally, RI and MX1; 200 Motorcycles: RGR, Rally; Motocross 250 NX, NE.

THE MX1

The most significant of these was the MX1, which was new at the show. Gilera's chief designer, Lucio Masut had the distinction of being the first man to address the issue of Italy's eighteen-month-old helmet law by relocating the MX1's fuel tank underneath the engine and up the right front corner of the full fairing, so that the filler was actually on the side of the bike. This had the effect of lowering the overall centre of gravity and maintaining handling characteristics as the content of the fuel tank became exhausted.

Masut had also designed a lockable space to be provided where the fuel tank would conventionally have been. This space was designed to take the rider's helmet and a waterproof oversuit, as well as such items as documents, or, alternatively, the pillion passenger's helmet if your girlfriend needed picking up from work, for example.

Although the MX1 was based on the successful KZ/KK series, it had been designed to provide what was essentially a new bike. The basic format was still the

Two-Strokes – the New Breed

From the KZ-KK series came the MX1. But, although based on the same concept, it was essentially a new bike. The basic format was still the same, with 56 × 50.6mm dimensions, carbon-fibre reeds, an electronic APTS exhaust power valve, gear-driven balancer shaft and six-speed box. But, besides modifications to the engine, the MX1 chassis was all-new, consisting of a steel twin-spar structure given the fancy title 'Twinbox' with alloy swinging arm and bolted-on subsection for the seat. Gilera launched the new bike at the Milan Show in November 1987.

same, with 56 × 50.6mm dimensions, carbon-fibre reeds, the electronic APTS exhaust power valve, gear-driven balance shaft and six-speed gearbox, which journalist Alan Cathcart rated 'one of the best changes I've ever had the pleasure of using'.

Modifications, including new cylinder porting and revised exhaust, had upped the power to 28bhp at 10,000rpm (but in practice the electronic ignition was fitted with a rev limiter which came in at 9,500rpm).

The MX1 chassis was new, consisting of a steel twin-spar structure given the fancy title of 'Twinbox', with alloy swinging arm and bolted-on sub-section for the seat. Strangely, Gilera didn't change the, by now discredited, 16in wheel size, even though Pirelli went to all the trouble of providing the new MT75 tubeless tyres especially for the MX1, and Grimeca one-off five-spoke wheels. However, the 35mm Marzocchi forks and Gilera's own 'Power Drive' rear suspension were extremely effective, offering a ride quality far in excess of what could, at that time, be expected from a mere 125. The 260mm (front) and 240mm (rear) Grimeca brakes remained as before, being more than equal and adequate for such a light bike.

The MX1's fairing had been designed by the styling house Pininfarina, and developed in that company's wind tunnel – to provide both good protection and a low drag factor. The exhaust system was manufactured for Gilera by Lafranconi.

Ultimately the MX1 didn't achieve quite as many sales as Gilera had hoped, many being taken away by the Cagiva Freccia – itself the forerunner of today's Mito. In addition, Aprilia was by then becoming an ever stronger force.

Everyone was following the antics of the three companies involved in a frantic battle for supremacy in the all-important, Italian 125cc section of the market. The likes of Laverda, Garelli and Fantic already having had to admit defeat.

Two-Strokes – the New Breed

(Left) *On the 5 June 1988, a pair of MX1s broke a series of World Speed records in the 125cc category. These ranged from six to twenty-four hours in duration.*

(Below) *Team members, including riders and back-up personnel, with the two MX1s that succeeded in breaking a series of work records at the Nardo circuit in June 1988.*

WORLD SPEED RECORDS

In their efforts to gain a definitive lead over their opponents, Gilera joined with the Italian journal *Moto Sprint*, providing two MX1s that, on 5 June 1988, broke a series of World Speed Records in the 125cc category. These took place at Nardo and saw the Gilera/Moto Sprint team take honours in the Six-Hour, Twelve-Hour, Twenty-Four-Hour and 1,000km sections.

In the Six-Hour and 1,000km the riders were Corrado Curti, Carlo Lotti and Luigi Rivola. These three were then joined by Claudio Braglia, Silvio Rocco, Riccardo Cusi and Dario Ballardini.

All four distances had been held since May 1979 by a team of Honda riders.

By 1990 Gilera had again virtually changed their two-stroke model line, and it now comprised the following: Trend (50), Bullit (50), RC Top Rally (50), MXR (125), XR2 Marathon (125), RC Top Rally (125) and SP02 (125).

145

Two-Strokes – the New Breed

MORE NEW MODELS

The MXR was essentially an update of the MX1, the SP02 a Sport Production (racing) version with 5bhp more, lightweight wheels, a 300mm front disc, upside down forks, a smaller 2.85imp. gal (13ltr) fuel tank (MXR 3.7imp. gal (17ltr)) and the looks of a GP racer. Gilera claimed 282bhp per litre for the SP02's engine output potential.

Then, at the Milan Show in November 1991, Gilera produced another full pack of newcomers: Sioux (50), 503 (50), Apache (125), Crono (125), Freestyle (125) and CX (125).

Running along parallel with its sports roadsters was a lineage of up-to-the-minute dirt-styled bikes. First came the XR1 in 1987.

Next came the XR2 Marathon of 1989. Its specification included: 38mm upside-down front forks, Gilera 'Power Drive' rear suspension with single gas shock, maximum power of 29bhp at 9,750rpm, front and rear disc brakes, 3imp. gal (16ltr) fuel tank capacity with a reserve capacity of 0.5imp. gal (2.5ltr). Gilera claimed a top speed of 94mph (151km/h).

Two-Strokes – the New Breed

SP Crona 125 (1990–1993)	
Engine:	Liquid-cooled, single-cylinder, two-stroke with reed valve induction
Bore:	56mm
Stroke:	50.5mm
Displacement:	124.38cc
Compression ratio:	12.5:1
Maximum power (at crank):	Estimated, 30bhp
Lubrication:	Oil pump and separate oil tank
Ignition:	Electronic
Fuel system:	Dell'Orto 32mm carburettor
Primary drive:	Gears
Final drive:	Chain
Gearbox:	Six-speed
Frame:	Delta-box type, aluminium alloy
Front suspension:	Inverted telescopic fork, 40mm diameter stanchion, 130mm stroke
Rear suspension:	Vertical single-shock, square section aluminium swinging arm
Front brake:	Single, fully-floating 300mm disc with four-piston caliper
Rear brake:	Single 240mm disc, with two-piston caliper
Wheels:	Front 16in; Rear 17in
Tyres:	Front 100/80 × 16; Rear 130/70 × 17
Wheelbase:	53in (1,350mm)
Dry weight:	255lb (116kg)
Fuel tank capacity:	2.8imp gal (13.1ltr)
Top speed:	100mph (160km/h)

The Sioux and Apache were the latest versions of the RC Top Rally models; the 503 a tiny racer replica; the Crono the latest incarnation of the SP; the Freestyle was basically a Nordwest styling job on the 125 two-stroke and the CX was – well, like nothing anyone had ever seen before ...

THE FUTURISTIC CX

Some of these models had actually débuted in Italy that year, including the Crono and the CX.

The CX went on sale in Italy during May 1991, with an appearance that made everyone instantly run to see what it was. It had first appeared in the metal at the Bologna Show in 1990 and was a production version of a 'dream bike' creation which Gilera had displayed at Milan in 1989, complete with Monotube front suspension and a single-sided swinging arm.

What the CX offered was an alternative to the ever-growing power available from Italian 125 sports bikes – which severely questioned the law that allowed sixteen-year-olds to ride such bikes in totally unrestricted form. Instead, Gilera thought it could break this cycle by offering *avant-garde* design as an alternative choice to absolute power for youngsters looking to choose a certain kind of bike.

The idea, spawned by Gilera's Frederico Martini, was that, by using such an *avant-garde* machine, with its technology clothed in an equally distinctive set of designer

Two-Strokes – the New Breed

Brainchild of the former Bimota and Ducati engineer Frederico Martini, the CX125 was intended as just the beginning of an entire new family of Gilera motorcycles.

clothes, the chain of performance as the main criteria would be broken. Gilera saw the CX concept as *gran turismo* and something which could, at a later stage, be employed with larger capacity engines.

The futuristic CX. The 125 was intended as only the first of a new breed of Gileras. Four-stroke versions using the Dakota-series engine were foreseen but not produced. Even today it looks ahead of its time, almost a decade after it went on sale in 1991.

Ing. Martini had joined Gilera from Bimota where he had been chief engineer, and in fact it was this same man who had designed the controversial Tesi.

Two months after joining Gilera, suspension specialists Piaoli had been commissioned to come up with a radial front suspension strut to Gilera's (Martini's) specification. Contrary to the expectations of many, who assumed the CX shown at Milan that November was merely an attention-grabbing design exercise, a little over a year later the machine was actually in production.

It employed a slightly modified version of the SP02's twin-spar, alloy look-alike steel chassis with single-sided swinging arm and single-strut front end, with what

Gilera termed 'Single Suspension System' (SSS) but everyone else called Monotube.

The CX front end was *not* a hub-centre as on the Tesi or ELF, but effectively one half of a modern, upside-down telescopic fork. As such it came close to being a production version of the Dutch White Power company's Monofork, which had been under development for the previous two years.

The CX without its clothes. Basically it used an SP02-derived chassis and detuned engine (but still good for 100mph! (160km/h)), with entirely new front end and wheels. Designer Frederico Martini (ex Bimota) saw it as a more sensible approach than the outright 'speedwars' of the racer-replica design concept. But the CX did little to curb Gilera's ailing fortunes.

The CX front suspension was still a fork, complete with pivoting steering head, fork yokes and hydraulic damping inside the suspension strut – only a fork rationalized to the ultimate degree.

Compared to the upside-down forks fitted to the SP02/Crono race-replica street bikes, here was a 13 per cent reduction in weight, much of it unsprung, accompanied by no less than a 40 per cent increase in stiffness, reduced gyroscopic effect, central location along the longitudinal axis of both front and rear discs, giving better-balanced braking, more space around the steering head, more direct steering input, lighter steering action, and easy replacement of the front wheel, as well, of course, as the rear with its single-sided swinging arm.

Martini was quoted as saying:

> The design allows greater stylistic freedom, not only aesthetically but also aerodynamically: SSS allows us to lower the steering head to reduce frontal area, and the design bodywork with a low coefficient of drag (hence the CX name), which improves rider comfort and maximum speed, as well as reducing fuel consumption.

Journalist Alan Cathcart put all the theory to the test during the spring of 1991 when he tested a CX over the old Monza bankings. This area of Monza had hardly been used for twenty years and Cathcart found the old surface 'incredibly bumpy'. Also revealing:

> In spite of its 264lb (119kg) feather-weight build and the relatively short travel of the Paioli-sprung suspension – 100mm at the front, 135mm at the rear – the CX was surprisingly stable and even comfortable. The rising-rate rear end felt especially responsive and gives the ride of a bigger bike, in turn reinforcing the aura of sophistication the bike gives off when you look or sit on it: for all its relatively diminutive size, this is a motorcycle with presence.

A notable difference was also the use of 17in wheels both front and rear (the 1991 Crono had at last gained a 17in for the rear, but retained the smaller wheel size at the front).

Martini planned the CX as the first of a Gilera family of similar unconventional design. But the big question remained – was the market ready for them? What would happen if Italy's notoriously fickle teenagers rejected the CX in favour of the latest neo-GP street racer? Martini was quoted as saying:

> We will pursue the CX concept whatever happens, because I don't see it so much appealing to our existing market, as to people who would like to have a bike but are frightened to buy one because they don't think they're expert enough. Face it – most bikes on the market today are threatening: either they're racing replicas which make you feel you have to be Eddie Lawson to ride them properly, or enduro bikes that demand the skills of Peter Hansel or Eddi Orioli. I want to create bikes which, far from frightening people away from motorcycling, encourage them to take it up, but which aren't so boring that they don't convey any excitement.
>
> Instead of fear, I want my bikes to offer satisfaction. The reason bike sales are dropping all around the world, and especially in Japan, is that people are intimidated by modern motorcycles, either by their aggressive appearance or technical complexity, with machines like the CX, and also the Nordwest. I want to build bikes for people mature enough to enjoy the essence of motorcycling, not the image alone.

THE GFR SPORTSTER

All fine words and, no doubt, well intentioned, but the fact was that potential customers wanted sports bikes, and, as if to finally admit this point, Gilera launched their 1993 model range, headed by a new outright sports 125, the GFR. This was a restyled, remodelled Crono. In fact its styling closely followed that of the extremely successful Aprilia Extrema.

The GFR at 260lb (118kg) (dry) was actually slightly heavier than the

Two-Strokes – the New Breed

(Left) *There is no disputing that the CX concept was futuristic, but motorcyclists are notoriously conservative.*

(Right) *Equally exciting was the Crono 125 SP02. First seen at the end of 1991, it was right up there in the speed stakes with the best, which by this time meant Cagiva and Aprilia.*

Crono. But in its favour was 17in wheels both front and rear (at last!), a higher compression ratio (13 instead of 12.5:1), a larger 320mm front disc (previously 300mm) and improved cylinder porting. It shared the same Dell'Orto VH SB 32 carb, premix lubrication, six-speed gearbox, electronic CDI ignition, 12v electrics and 2.85imp. gal (13ltr) fuel capacity.

Another new-for-'93 model was the 125R, a state-of-the-art trail mount. This used the same basic engine as the GFR sportster, including the 13:1 compression ratio and 32mm carb. It really looked the part, and its bright red/purple/white paint job (including a white frame) made it a real head-turner.

PROBLEMS, PROBLEMS

But all this rapid progress and never-ending stream of new bikes caused its own problems. These included obsolete models within too short a period, overworked and over-stretched parts back-up, excessive development costs, and even then their main rivals in the two-stroke field – Aprilia and Cagiva – still sold more bikes during the early 1990s. Gilera was in trouble once more ...

As outlined in Chapter 10, Gilera tried to solve some of their problems with the possibility of joint ventures with Honda, but in reality this plan was to be stillborn, and with it the hopes of the Gilera marque.

By the autumn of 1993 Piaggio had finally had enough. It couldn't see an end to Gilera's mounting losses and so pulled the plug.

Martini and his team had done their best, but sometimes even this is not enough. Perhaps their biggest mistake was to try and create new customers and not secure their existing ones. For Martini it was a lesson he should have heeded first time around with the Tesi – which almost destroyed Bimota.

9 Four-Strokes – the New Breed

THE DAKOTA

The first the world saw of the new Gilera four-stroke during the 1980s was at the biennial Milan Show in November 1985. Not only did the famous old Arcore factory have one of the biggest and most impressive stands at the exhibition, but also an exciting new thumper in the shape of the 350 Dakota trailster, with the added promise of a larger version in the pipeline.

And Gilera's first new four-stroke in over a decade showed that it wasn't just two-stroke models which were receiving the modern treatment in the Arcore R & D shop.

FOUR-VALVES

Designed by Ing. Lucio Masut, the newcomer employed every trick in the book – including liquid cooling, four-valve cylinder head, twin exhaust ports, twin 25 Dell'Orto carbs, double overhead camshafts driven by toothed belt, a balancer shaft (gear driven directly off the crankshaft) multi-plate hydraulically operated

Designed by Ing. Lucio Masut, the Gilera 350 Dakota was introduced unexpectedly at the 1985 Milan Show. It was Gilera's first really new four-stroke for over a decade and showed that the Arcore company was still capable of building world-class machinery.

Four-Strokes – the New Breed

350 Dakota (1986–1989)	
Engine:	Liquid-cooled, dohc single-cylinder, four-stroke with four-valves
Bore:	80mm
Stroke:	69.4mm
Displacement:	348.89cc
Compression ratio:	9.5:1
Maximum power (at crank):	39bhp (33bhp at rear wheel)
Lubrication:	Wet sump, gear pump
Ignition:	Electronic capacitor discharge (CDI)
Fuel system:	2 × Dell'Orto 25mm carburettors
Primary drive:	Gears
Final drive:	Chain
Gearbox:	Five-speed
Frame:	Tubular steel, with engine as stressed member
Front suspension:	Telescopic fork, Marzocchi Motocross type, 38mm diameter stanchion, 220mm stroke
Rear suspension:	Square section swinging arm, single Marzocchi shock absorber; 220mm travel
Front brake:	Single 260mm disc with Brembo two-piston calipers
Rear brake:	Drum 160mm
Wheels:	16in front, 18in rear
Tyres:	Front 3.00 × 21; Rear 4.60 × 17
Wheelbase:	57in (1,450mm)
Dry weight:	326lb (148kg)
Fuel tank capacity:	4.8imp. gal (22ltr)
Top speed:	90mph (145km/h)

wet clutch, five-speed gearbox and Japanese electronic ignition and electric starter. There was also a forged piston and a one-piece crankshaft which ran on anti-vibration ring bearings. The pity was that all this hi-tech only produced 33bhp at 7,500rpm from the 348.8cc (80 × 69mm) mill. But at least the 500 (which took another eighteen months) promised a 25 per cent power increase with no additional weight to speak of.

DUAL-PURPOSE

Gilera had opted to build a dual-purpose, on-off-roader rather than a pure street bike because, at the time, sales of machines like Yamaha's Tenere were riding high. The Paris-Dakar style was all the rage both in Italy and around Europe; in fact everywhere except Great Britain! Even though the Dakota's power output was disappointing in view of its technical gizmos, the in-house Gilera styling job was generally accepted as being superb. As proof of this, author Alan Cathcart picked the Dakota to feature as only one of two machines with engines of less than 750-class in his 1988 book, *Dream Bikes*.

But besides the lack of power, the wide 5imp. gal (23ltr) fuel tank and sculpted bodywork was slammed by serious off-road riders for its enforced 'knees-out' riding stance which, unless the rider adopted a more rearward seating position, (which then altered the weight distribution) was extremely uncomfortable.

Four-Strokes – the New Breed

The Dakota engine employed every trick in the book – liquid cooling, four-valves, twin exhaust ports, twin carbs, dohc, a balance shaft, multi-plate hydraulic clutch, five-speeds, Japanese electronic ignition and electric starter.

The Brembo-sourced hydraulic clutch was a very neat design, there was also a back-up kick-starter to complement the electric button. The oil filter cartridge is also visible in the bottom right-hand corner of this photograph.

THE ER VARIANT

Gilera responded in a positive fashion by introducing the ER variant in 1987, just as the 500 model 492cc (92 × 74mm) came into production. The ER was targeted firmly at the committed green lane rider, with its smaller 3.3imp. gal (15ltr) tank and the twin coolant radiators now shrouded by abbreviated plastic panels. These two changes enabled the driver to sit further forward in a more comfortable position. The original model was also retained and, with both versions available in 350 and 500 form, this meant that there were actually four different Dakota models to choose from.

With a dry weight of no less than 324lb (147kg) in both engine capacities, the Dakota was one of the heaviest machines in its class. This showed up most under heavy braking when the single 260mm disc and its Grimeca four-piston caliper was stretched to its limit to cope. The Gilera engine was the most sophisticated in the on-off-road sector of the market, but with its great weight and power-sapping ultra efficient (quiet!) exhaust system it was not really suitable for pure competition, even in its 500 form.

However, the engine was very strong and the square-tube enduro-type chassis was fully capable of taking more power. The result was that at the Milan Show in 1987 Gilera launched the XRT.

THE XRT

The XRT was everything the Dakota was not, with its 569cc (99 × 74mm) engine and low 150kg (331lb) weight (it was also built for the Italian market with the 348.8cc engine). Maximum power was now up to 47bhp at 7,000rpm and Gilera claimed 105mph (169km/h).

All the good features of the Dakota engine had been retained, although the hydraulically operated clutch had been abandoned in favour of a cable operated device.

Four-Strokes – the New Breed

The XRT 600 was a definite improvement over the 350/500 Dakota, with its more torquey 569cc (99 × 74mm) engine and superior power-to-weight ratio. There was also an Italian-market 350 version.

admired the Dakota, the XRT just looked that much better. It was also a better bike where it counted, out on the street or trail, with its improved performance, superior power-to-weight ratio, better braking and improved riding position. Viewed from every angle it was an excellent motorcycle.

JAPANESE INVOLVEMENT

Also at the 1987 Milan Show was a sports roadster with the engine taken from the larger Dakota. This was not a bike conceived by the Italians themselves, rather one which had been created especially for the single-cylinder-mad Japanese market, and specially commissioned by the huge Japanese trading house C. Itoh.

Although the frame remained essentially the same, there was a new, stronger swinging-arm. The rear drum brake of the Dakota had been replaced by a 230mm disc, with the original 260mm disc up front retained. But with the benefit of a more efficient rear stopper, braking was noticeably improved.

The 331lb (150kg) weight was for the model without electric start. The same 180w alternator was used for both the electric- and non-electric-start versions of the XRT, but they came with nine and fourteen-amp hour batteries respectively.

In the styling department the XRT was a generation ahead of the Dakota, even though a mere two years before the latter had been a real head-turner. The XRT simply looked stunning with its flowing bodywork, upswept single silencer, twin headlamp fairing, and massive sump shield/fairing enveloping the bottom end of the engine. However much you might have

One of the surprises of the 1987 Milan Show, the Nouvo *Saturno was the result of Ital-Japanese co-operation between Gilera and the trading company C Itoh of Tokyo. If all this seems strange one has to appreciate that, at the time, the Japanese were madly devoted to both classic bikes and single-cylinders, hence the reborn Saturno.*

155

Four-Strokes – the New Breed

Piuma Racer

The 1987 Milan Show had been the launch pad for the Nuovo Saturno *Bialbero* (double overhead cam). The engine had been based on the liquid-cooled four-valve unit first seen in the 350 Dakota single enduro-styled machine which had débuted two years earlier at Milan in November 1985.

This series of engines was largely the work of Gilera engineer Ing. Sandro Columbo. And, although originally conceived as a joint venture with the Japanese trading company C. Itoh, the new Saturno was soon to find its way onto the European race circuit, thanks to the popularity of mono- (single) cylinder racing events. However, instead of simply tuning the existing 491cc (92 × 74mm) unit, the racing version employed the still larger displacement 569cc (99 × 74mm) assembly found in the XRT Paris-Dakar styled mount which Gilera had begun selling during the 1998 model year.

During the early 1990s, Gilera created a new offshoot to oversee its sporting activities, Gilera Corsa (racing). This was to oversee the design and manufacture of both on- and off-road competition machines. By the 1991 Milan Show the first examples of the racing Saturno, the Piuma (Puma) were ready. Besides the production version, there was also a very limited number of larger capacity 620cc versions. These latter machines were only made available to factory-supported riders in the European Super Mono race series with rounds staged as support events at World Superbike (WSB) meetings.

Brand new Nuovo Saturno Piuma *(Puma) racer on display at the 1991 Milan Show. It was a factory effort, intended to compete in the European Supermono (Singles) road-racing Championship series.*

One of the Saturno Piuma racers fitted with lightweight carbon-fibre bodywork (fairing, seat and front mudguard).

Piuma Racer

But, right from the start, the factory 620cc and production 569cc models were to be blighted by engine vibration problems; another major problem was crankcase failure – in truth, one probably led to the other.

Track Test
I was able to sample a virtually new Saturno Piuma, kindly loaned to me by Gilera enthusiast Sally Kelly at Mallory Park on 12 July 1998. Certainly the machine's best feature was its superb looks – even better in real life than in photographs. Vibration was most definitely its most serious shortcoming; okay, most singles vibrate, but on the Piuma this was more than just a nuisance – it almost ruined the riding enjoyment. Almost, but not quite, as the excellent handling, braking and general feel of the machine (vibes excluded) made up for its shortcomings.

The author with Sally Kelly's freshly delivered Nuovo Saturno Piuma at Mallory Park, 12 July 1998. Looks killed, but vibes were a problem.

THE NUOVO SATURNO

In fact this prototype was then flown to Japan and displayed by the Itoh organization at the 'Mega-Show' which began in Tokyo on 20 December. The *Nuovo* (New) Saturno project was very much the work of the two men, Gilera engineer Sandro Colombo and the Japanese technician, N Hagiwara. The general idea was to recreate the Saturno concept, but in a modern guise.

TECHNICAL DETAILS

Using the 492cc Dakota engine, the Ital-Japanese pairing created a compact café racer (it was also built with the smaller engine). Weighing in at 296lb (135kg) the Nuovo Saturno employed a one-off trellis steel-tube frame and an alloy swinging arm with eccentric adjustment for the final drive chain. No expense had been spared in selecting the best quality components; the footrests, rear brake and gear change levers were all in aluminium, as was the kick-starter.

The 300mm floating brake disc at the front was operated by a four-pot Brembo 'Gold-line' caliper, with a 240mm disc and dual piston caliper at the rear, while the Marvic cast alloy wheels each had three hollow spokes. Both wheels were 17in diameter with 110/70 front and 140/70 rear tyres of VR rating.

The 40mm front forks had 120mm of travel and were of conventional design. The rear end was taken care of by a single shock with 130mm travel. The shock was of the racing type with multi-adjustment. Other details of the machine included a half-fairing, clip-on handlebars, indicators, twin mirrors (fairing mounted), Japanese-

Nuovo Saturno (1988–1991)	
Engine:	Liquid-cooled, dohc single-cylinder, four-stroke with four valves
Bore:	92mm
Stroke:	74mm
Displacement:	492cc
Compression ratio:	9.5:1
Maximum power (at crank):	45.1bhp (36.5bhp at rear wheel)
Lubrication:	Wet sump, gear pump
Ignition:	Electronic capacitor discharge (CDI)
Fuel system:	2 × Dell'Orto PHM40VS (PHM36PS)
Primary drive:	Gears
Final drive:	Chain
Gearbox:	Five-speed
Frame:	Tubular steel, space triangulated. Engine as stressed member
Front suspension:	Telescopic fork, Marzocchi
Rear suspension:	Swinging arm, single Marzocchi shock absorber
Front brake:	Single 300mm disc with Brembo four-piston caliper
Rear brake:	Single 240mm disc with Brembo two-piston caliper
Wheels:	17in front and rear
Tyres:	Front 110/70 VR17 Pirelli MP7S; Rear 140/70 VR17 Pirelli MP75
Wheelbase:	55.5in (1,410mm)
Dry weight:	308.5lb (140kg)
Fuel tank capacity:	4.4imp. gal (20ltr)
Top speed:	111mph (178.5km/h)

style bar-end weights, a single seat and racing-style tank, a plastic chainguard and a rear hugger mudguard in the same material. The exhaust system, finished overall in black, was a siamezed exhaust pipe exiting into a single silencer. This was mounted just below the line of the seat base on the offside of the machine. In typical café-racer fashion there were the bare minimum of fitments to the bike, typified by no centre stand, simply a side (jiffy) one. The finish of the whole machine was in Italian racing red.

As the engine literally hung in the frame it was readily accessible, unlike the majority of modern sports bikes with their total enclosure.

FOR GENERAL SALE

Although everyone believed that the Saturno was a Japan-only bike, this proved to be wrong and by the end of 1988 it went on general sale.

In the United Kingdom, Piaggio's representative at that time was the Heron Group, based in Crawley, West Sussex. Its subsidiary, Vespa UK, spent over two years deciding whether to actually import the Saturno (and other four-stroke Gileras). Then, in December 1990, a small batch of machines was imported, priced at £4,999. This, in the author's opinion, was too high a price (at that time a CBR600 Honda was less than £4,000!), but, even so, all the machines brought in were sold, leading

Four-Strokes – the New Breed

The Lady Loves...

In this case it is, or was, an immaculate Nuovo Saturno.

Ann Cheetham is a dedicated bike enthusiast whose working week is spent as a social care assistant at Chesterfield Hospital.

She began her motorcycling career on an aged BSA Bantam two-stroke, the only claim to fame of this particular machine was its ability to seize up almost daily! After passing her motorcycle test in 1975, Ann rode a Honda CB400F for several years before acquiring the Nuovo Saturno in September 1997. She took her new acquisition to Scarborough for 'Classic Week', the following day.

Originally Ann and her husband Peter had visited the Saturno's then owner, John Ownsworth (an ex-racer from Barnsley, South Yorkshire), partly because they had never seen one in the metal before. It was love at first sight; John had kept the machine in a specially made, centrally heated wooden crate – and it was thus truly immaculate.

Ann found the Gilera 'low and light', its 36bhp-300lb (136kg) power-to-weight ratio proving ideal. Track days followed at Cadwell Park, Donington Park, and Oulton Park. At the latter venue she was guided round the circuit by a classic racer (750 Seeley-Commando) who was suitably impressed enough to recommend that she 'race rather than just do the track days'.

But disaster was just around the corner and her racing career on the Gilera never got underway before the bike was stolen from the family's garage – with all the riding gear – in November 1998.

Nothing was heard of the Saturno until August 1999, when it was seized during a police raid at an address in Mansfield (some twenty miles (thirty-two km) distant). When Ann and Peter went to identify the bike at Sutton-in-Ashfield police station, it was, to quote, 'trashed and incomplete – I shed more than a few tears'.

Instead, the initial racing season was completed on a 350K4 Honda and, despite a couple of spills along the way, Ann is campaigning on a 250 Ducati in 2000. But there will always be a special place in her heart for a certain bright red Gilera ...

Ann with her beloved Nuovo Saturno, Gary Walker Memorial Meeting, Mallory Park, 12 July 1998.

Ann Cheetham, classic racer and general bike enthusiast.

Four-Strokes – the New Breed

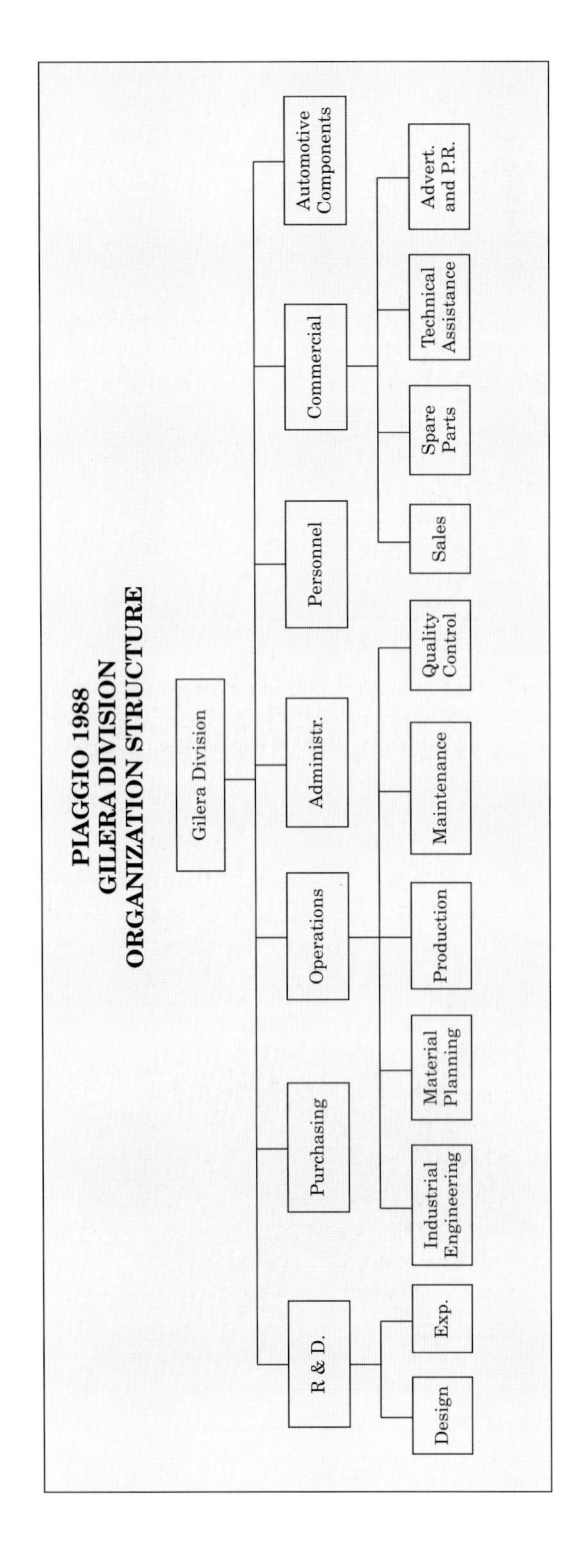

PIAGGIO 1988 GILERA DIVISION ORGANIZATION STRUCTURE

```
Gilera Division
├── R & D
│   ├── Design
│   └── Exp.
├── Purchasing
│   ├── Industrial Engineering
│   └── Material Planning
├── Operations
│   └── Production
├── Administr.
│   ├── Maintenance
│   └── Quality Control
├── Personnel
│   └── Sales
├── Commercial
│   ├── Spare Parts
│   ├── Technical Assistance
│   └── Advert. and P.R.
└── Automotive Components
```

Former Gilera works rider and World Champion, Geoff Duke posing with a 500 Saturno in 1990, one of the first batch of fifty machines imported into Britain by the Heron Corporation.

A Nuovo Saturno being put through its paces at an Owners' Club track day at Cadwell Park, circa 1998; rider Peter Cheetham.

Four-Strokes – the New Breed

Factory entered RC600Rs dominated the 600cc category of the Paris-Dakar Rally during the 1990s. This is the customer version; it shared its basic dohc four-valve engine with the Nordwest and standard RC600, but power was up to 54.56bhp at 7,500rpm. This was a serious dirt iron with the very best equipment – Excel alloy rims, 46mm heavyweight forks, dual rear wheel sprocket, Tripmaster speedo and comprehensive sump bash-plate.

the way to more Saturno imports and other models such as the XRT and its successors.

THE RC600 AND NORDWEST

Throughout the late 1980s Gilera took a serious interest in the Paris-Dakar and similar rallies. This culminated in 1990 with a Gilera class victory in the Paris–Dakar event.

An offshoot of this was a plethora of production models, including the RC600, RC600R and Nordwest.

All used a 558cc (98 × 74mm) variant of the motor used on the XRT. Quite why Gilera chose to decrease the bore size by 1mm is unclear, but this had the effect of reducing the capacity from 569 to 558cc. There were other differences too, the carburettor size being increased from the 25mm instruments on the XRT (the same as the Dakar series) to 30mm on these later models.

The gear ratios remained the same but outright power was significantly improved – 53bhp on the RC600 and Nordwest and 54.5bhp on the RC600R – all three machines peaking at 7,500rpm.

Works-entered RC600Rs dominated the 1991 and 1992 600cc-class of the Paris-Dakar, making it one of the truly great machines in the history of the event (others including the BMW GS, the Cagiva Elefant and a small number of exotic Japanese factory entries.

NORDWEST PROVES A WINNER

But it was the Nordwest which really caused a stir with its unique style and beautifully flowing lines.

The Nordwest boasted an uprated specification (at least for street use) over the RC600, with its three-spoke cast alloy wheels, twin 270mm semi-floating discs (front) and single 240mm disc (rear) and both wheels shod with Michelin Hi-Sport 17in tyres (120/60 front, 120/70

Four-Strokes – the New Breed

RC600R (1990–1993)	
Engine:	Liquid-cooled, dohc, single-cylinder with four-valves and twin exhaust ports
Bore:	98mm
Stroke:	74mm
Displacement:	558cc
Compression ratio:	10.5:1
Maximum power (at crank):	54.5bhp @ 7,500rpm
Lubrication:	Wet sump
Ignition:	Electronic
Fuel system:	2 × 30mm TK carburettors
Primary drive:	Gears
Final drive:	Chain
Gearbox:	Five-speed
Frame:	Steel, tubular
Front suspension:	Telescopic fork, 46mm stanchion diameter, 290mm stroke
Rear suspension:	Single shock, 295mm stroke
Front brake:	Single disc, 260mm
Rear brake:	Single disc, 220mm
Wheels:	Front 21in; Rear 18in
Tyres:	Front 90/90 × 21VR; Rear 120/90 × 18
Wheelbase:	58in (1,480mm)
Dry weight:	304lb (138kg)
Fuel tank capacity:	2.64imp. gal (12ltr)
Top speed:	100mph (160km/h)

The Nordwest was Gilera's best selling four-stroke during 1992 and 1993. Although it was something of a styling statement, it was nonetheless a fine motorcycle. The spec included three-spoke alloy wheels, twin 270mm front discs/single 240mm rear disc, Michelin Hi-Sport tyres, inverted forks and Gilera's own 'Power Drive' single rear shock.

Four-Strokes – the New Breed

	Nordwest (1992–1993)
Engine:	Liquid-cooled, dohc single-cylinder, four-stroke with four valves and gear-driven balance shaft
Bore:	98mm
Stroke:	74mm
Displacement:	558cc
Compression ratio:	10.5:1
Maximum power (at crank):	53bhp @ 7,500rpm (41bhp at rear wheel)
Lubrication:	Wet sump
Ignition:	Electronic capacitor discharge (CDI)
Fuel system:	2 × TK 30mm carburettors; 1 CV type, 1 slide type
Primary drive:	Gears
Final drive:	Chain
Gearbox:	Five-speed
Frame:	Spine, steel, with engine as stressed member. Bolted-up sub-frame
Front suspension:	Telescopic, inverted Paioli, 40mm diameter stanchion, 160mm stroke
Rear suspension:	Aluminium swinging arm, rising-rate monoshock, 150mm travel
Front brake:	Twin floating 270mm discs, with four-piston calipers
Rear brake:	Single 240mm disc, with two-piston caliper
Wheels:	17in front and rear
Tyres:	Front 120/70 × 17 Pirelli MT60RR radial; Rear 160/60 × 17 Pirelli MT60RR radial
Wheelbase:	56in (1,420mm)
Dry weight:	308.5lb (140kg)
Fuel tank capacity:	2.6imp. gal (12ltr)
Top speed:	110mph (177km/h)

Note:
The Nordwest was fitted with an electric starter as standard equipment.

Racer Nigel Lee on his Nordwest at Silverstone during the mid-1990s. The model was surprisingly competitive against pukka racing machinery.

Four-Strokes – the New Breed

Gilera Network co-founder Pete Fisher aboard his Nordwest at Ramsey Hairpin, during the Lerghy Frissel Hill Climb, 1996.

rear). There were also upside-down 40mm (stanchion diameter) front forks and Gilera's 'Power Drive' rear suspension (single shock). Other features (some of which were shared by the RC600 series) included: an alloy swinging arm, speedo, tacho and temperature gauge, a cast aluminium rear carrier, stylish bodywork, oblong indicators, plastic chainguard and 520-size final drive chain. And, unlike the RC models, the Nordwest did not feature a back-up kick-starter. Gilera also had other four-strokes in the pipeline, but none made it into production.

10 End of the Road

In 1969, Gilera was taken over by the Piaggio group, who, by the beginning of the 1990s, were Europe's largest manufacturer of two-wheeled vehicles and the third largest in the world.

At the end of 1991 the Piaggio empire could boast an impressive array of companies, not just in the powered two-wheeler field, but also in engineering, trading, financial activities, bicycles and industrial products.

PIAGGIO VEHICLES

Closer to home the Piaggio Vehicles section comprised not only Gilera, but also Vespa and Puch; the latter purchased a few months before from Steyr-Daimler of Austria.

Piaggio also had no less than fourteen oversees licence holders, manufacturing mainly Vespa-originated models. These countries included Bangladesh, India, Indonesia, Iran, Syria, Nigeria, Pakistan, Communist China, Taiwan, Thailand and Tunisia. There were also Piaggio controlled importers in both France and Germany (and later in Britain). And in Spain the former licence holder had come under Piaggio control in 1986, as Moto Vespa SA, with its head office in Madrid.

In 1991 Gilera was a division of Piaggio & Co SpA, manufacturers not only of two-wheel vehicles, but also of automotive components (for example Gilera supplied companies such as Fiat with servo breakers, rocker arms, pedals, brake levers and pulleys).

The total numbers of employees at Gilera was 720, of which 590 were blue-collar workers. Those in two-wheel production numbered 508 and 402, respectively. Of these no less than sixty were employed in the R & D department.

Also in 1991 Piaggio had established

Winner of the 1991 Paris – Le Cap Rally, the works entered RC600 Gilera enduro seen on the company's stand at the Milan Show later that year.

End of the Road

Gilera *Corsa* (racing), to the tune of seven billion lire, to promote Gilera in racing. A factory, like Honda's HRC, was set up especially to design and build racing prototypes, headed by former Bimota design chief Frederico Martini, with Oliviero Cruciani appointed as race manager.

At the Milan Show in 1991 it was announced that Gilera would be contesting the 1992 250cc World Road Racing Championship series, its riders being Carlos Lavado (who had previously won no less than nineteen GPs in the 250 category) and Jean Phillipe Ruggia. The machine they would use was a brand new 75-degree V-twin two-stroke, really the same basic formula as its two main rivals in the class, Honda and Yamaha.

GFR250

Engine:	Two-stroke – two cylinders in 75-degree V single crankshaft with three main bearings
Cooling:	Liquid
Intake:	Segmented valves
Exhaust:	Electronic Gilera APTS valves
Bore & stroke:	56 × 50.7mm
Compression ratio:	16:1
Piston:	Single ring, in forged aluminium
Carburettors:	Two downdraft Dell'Orto units, 40mm diameter choke value, flat valve and electronic power jet
Ignition:	Digital CDI
Alternator:	Internal rotor Piaggio model
Gearbox:	Six-speed, extractable with electronic semi-automatic control device (This new device enabled the rider to change gear without having to choke the accelerator throttle and without touching the clutch)
Clutch:	Dry with sintered bronze plates
Radiator:	Aluminium with double passage over radiating pack
Exhaust:	Fully handcrafted steel exhaust pipes
Frame:	'Twin Box' boxed aluminium twin beam, carbon-fibre saddle support
Front suspension:	Kayaba fork with external adjustment of hydraulic stroke compression and extension titanium spring with externally adjustable preloading
Rear suspension:	Gilera 'Power Drive' system, with boxed aluminium fork and pin position adjustment, Kayaba damper with titanium spring and external adjustment of hydraulic stroke compression and extension
Front brake:	Two different Brembo discs: steel, 290mm diam. and carbon, 255mm diam., four-piston calipers and radial pump
Rear brake:	Steel Brembo disc, 190mm dia
Wheels:	Marchesini, magnesium alloy, front 3.50 × 17; rear 5.50 × 17
Tyres:	Michelin 17in front and rear
Fairing:	In combined carbon/kevlar
Fuel and lubricants:	Agip 3.5 per cent mixture
Weight:	209lb (95kg)
Wheelbase:	52in (1,320mm)
Variable steering angle:	22 degrees +/–2 degrees adjustable front travel
Fuel tank capacity:	4.8imp. gal (22ltr)
Maximum speed:	165mph (266km/h)

Jean Phillipe Ruggia, Gilera's big hope for success in the 250 Grand Prix when the company returned to the race circuit in 1992. Results were never to match expectations.

A more detailed look revealed twin Mikuni or Dell'Orto carbs (although it was ultimately planned to use fuel injection), reed valve induction with electronically timed exhaust valves, six-speed gearbox, balancer shaft and magnesium engine cases. The reed valves had been designed specially by the Austrian engineer Harold Bartol. A single crankshaft was used, this layout being employed to eliminate flexing at high rpm.

The chassis was in Delta-box aluminium, with Kyaba suspension front and rear. The upside-down front forks featured an adjustable steering head rake, and there was a fully progressive rear shock. The swinging arm was of the 'Banana' type, again in alloy. Both steel and carbon fibre were tested for brake-disc material, as were alloy and carbon fibre for the fuel tank – the latter not, however, being used due to an FIM ruling.

When I spoke to Frederico Martini at the 1991 Milan Show, he said that the 'great plan' would take some ten years and that 'Piaggio very much viewed Gilera as Fiat did Ferrari'.

1993 Gilera Racing Team Organization Chart

Team Manager:	Oliviero Cruciani
Riders:	Paolo Casoli (ITA)
	Alessandro Gramigni (ITA)
Engine Manager on the track:	Harald Bartol
Gilera Corse Manager on the track:	Romolo Ciancamerla
Casoli staff:	One chief mechanic
	Two assistant mechanics
Gramigni staff:	One chief mechanic
	Two assistant mechanics
Hospitality unit:	Vehicle equipped to provide 100sq m of hospitality space, with a restaurant and meeting rooms
Hospitality staff:	Four
Vehicles:	One towing vehicle with workshop
	Semi-trailer
	Camper
	One van for personal transfers
	One van for fast material transport
Sponsors:	Agip, Michelin
	Compagnucci, Dell'Orto, Domino Giannelli, M Roberts, NGK, Pisani, Catene Regina, Triom

End of the Road

The 1993 Gilera Riders	
Alessandro Gramigni	
Born:	Florence, 29.12.1968
Height:	1.72m
Weight:	63kg
Eyes:	Blue
Hair:	Light brown
1st Race:	1987
1st Grand Prix:	1988
Best Result:	1992 World Champion 125cc class
Hobbies:	Motorcross, windsurfing
Favourite colour:	Red
Lucky Number:	Thirty-nine
Paolo Casoli	
Born:	Castelnuovo Monti (RE), 18.8.1965
Height:	1.77m
Weight:	68kg
Eyes:	Brown
Hair:	Brown
1st Race:	1982 (125cc)
1st Grand Prix:	1986 (125cc)
Best Result:	1st Portuguese GP World Championship
Racing Classes:	125 – 250
Hobbies:	Music
Favourite Colour:	Yellow
Lucky Number:	Seven

DIFFICULT TIMES

For the next two years Gilera *Corsa* spent plenty of lire, but found results difficult to come by, even though its power output of 84–85bhp was the same as that generated by Honda's championship winning NSR.

Somehow the racing division couldn't quite come up with the goods, even though Gilera was a force to be reckoned with in off-road events such as the Paris-Dakar – the 600cc class being dominated during the same period by the company's RC600R.

PIAGGIO UK

At the end of 1982 a wholly-owned UK subsidiary, Piaggio Ltd, was set up. This

The class-winning RC600, Paris-Dakar Rally, 1991. During the early 1990s the factory had great success in this, the toughest of all motorcycle events.

took over from the Heron Corporation's Vespa UK which had been handling imports of both Vespa and Gilera up to that time. Of course, it is also worth noting that Heron had run into financial problems of their own, with chairman Gerald Ronson falling foul of the British legal system through rogue share dealings, in what was to become known as the Guinness Affair. The Piaggio-owned Gilera UK had as its managing director Giuseppe Tranchina and, under his regime, sales started to pick up, with the 600 Nordwest in particular becoming quite a cult machine during 1992.

THE HONDA CONNECTION

Then, in March 1993, came what many considered at the time to be a historic moment, the announcement that an agreement had been signed in Tokyo the previous month between Gilera and Honda.

As *Bike* magazine reported: 'Honda, the world's largest motorcycle manufacturer and Piaggio, Europe's two-wheeled leader, are joining forces in a move which could have massive repercussions on world motorcycling.'

Bike's headline blazed 'Honda/Gilera Link-up Fire-Blade-powered Gilera maybe?' *Bike* also said that the agreement compared with the Honda car division's tie-up with Rover. (This was before BMW's takeover of the Rover Group.)

Actually, the press, in their rush to grab a headline-popping story, missed one vital point. The document signed in Tokyo was a letter of intent, *not* a binding agreement. What this letter of intent said was that Honda and Gilera would: 'attempt to develop a five-year plan of European market technology transfer'.

The main points were as follows: Honda would supply engines and key components for a new range of Gilera superbikes ranging from 600 to 1,000cc, due to hit the streets in 1995. Honda and Piaggio (which for the purposes of the proposed agreement meant Gilera and Vespa) would strive to standardize parts between the marques to reduce costs. Honda and Piaggio would work to establish joint development of engines for small and medium-sized bikes.

The 1993 Gilera 250GFR Grand Prix machine. A 75-degree 249.9cc (56 × 50.7mm) V-twin with reed valve induction and exhaust valves. The gearbox was a six-speeder.

Postcard showing Alessandro Gramigni with the 1993 Gilera GFR250 V-twin.

End of the Road

In essence, the proposed agreement sought to follow a joint venture that was already up-and-running between Piaggio and Daihatsu in the field of light commercial vehicles. Piaggio president Gustavo Denegri expressed his desire to 'develop Piaggio's motorcycle division under the Gilera brand name into a major world force'.

END OF ARCORE

All brave words but, in truth, the Gilera re-birth was constructed on shifting sand rather than bricks and mortar.

Unbelievably, a mere eight months after the Tokyo signing ceremony, and on the very eve of the Milan Show in November 1993, Piaggio announced to a stunned world that it was closing the Gilera arm and with it the famous old Arcore works.

Why then such a rapid turnabout from the great optimism of a few months earlier?

There was only one answer: Gilera was proving an unacceptable drain on Piaggio's resources, and the Group's financial mandarins could see no light at the end of the tunnel. It was the end of the road for the eighty-three-year-old marque, which the young Giuseppe Gilera had founded back in 1909.

One of the final Gilera models to be built at the Arcore factory before its closure at the end of 1993, the Bullit was powered by a 49cc reed-valve, two-stroke engine and boasted a frame design which would have been fully capable of gracing a much larger and more powerful machine.

A sad sight – the rear of the famous Gilera factory at Arcore in October 1994, a year after motorcycle production had ceased at the plant.

The CX Project

Name the most futuristic motorcycle of the 1990s: some might say the Bimota Tesi, others the Yamaha GTS, but in many ways the title should go to the Gilera CX.

The CX was 'only' a 125 street bike, but, had Gilera continued manufacturing motorcycles after autumn 1993, its technology would have no doubt been carried through to the range of larger capacity four-stroke models.

If you discount the Bimota and Yamaha, which were never more than small batch exercises (in the Yamaha's case it was very much a giant sales failure), the Gilera CX125 was the first bike of the modern era to be fitted with something other than a telescopic fork, which subsequently reached series production for the street.

Although many observers had considered Honda would do this with the development work carried out by ELF in the 1980s, it was, in fact, Gilera who were to come up with the goods. The production version of the CX125 was first displayed at the Bologna Show in December 1990 – a year after exhibiting the prototype at Milan in November 1989.

The CX project was intended to embrace a whole family of new Gilera models for the mid-1990s and beyond. Instead, only the CX125 was actually put into production before the Arcore plant closed.

> **The CX Project** *(continued)*
>
> The CX entered production at the end of 1990 and featured both *mono tubo* (single-sided fork) front suspension (Gilera called it SSS – Single Suspension System) and single-sided rear swinging arm.
>
> Sales in Italy began in April 1991 and the first target was the domestic youth market, in what Alan Cathcart writing in the April 1991 issue of *Motor Cycle International* described as:
>
>> A laudable attempt to break out of the power and performance race which that nation's manufacturers have been involved in of late, resulting in 105mph race-replicas like the Cagiva Mito and Gilera's own SPO2 that have tested the law, allowing sixteen-year-olds to ride such bikes to the limit.
>
> The CX was designed by Ing. Frederico Martini, formerly of Bimota and the man largely behind the Tesi project.
>
> But did the CX chassis/suspension work and how did it compare to the Tesi? Testing a new example back in 1992, I found the CX125 to be a brilliant little bike – not just in the handling department, but also from a comfort point of view.
>
> For a start it is not a hub-centre type as on the Tesi, ELF or the Yamaha 1000GTS, but effectively represented one-half of a modern, state-of-the-art inverted telescopic fork. In many ways it was virtually a production version of the Dutch White Power Monofork which was developed at the same time. 'But why', I asked designer Martini 'haven't you employed a Tesi-type, hub-centre design, is not Gilera's SSS a compromise which, as such, may well retain some of the defects associated with the telescopic type?'
>
> Martini's reply showed that he still thought the Tesi's fork superior, but impractical for mass production on cost grounds if nothing else. And that, Martini reasoned, was why Honda hadn't employed the ELF design on a production roadster.
>
> Instead Martini and his team at Gilera chose to develop an entirely new fork in conjunction with suspension specialists, Paioli. The legalities were that, although Gilera owned the rights, they agreed to allow Paioli (who actually built the assemblies) to sell it to other manufacturers from 1994 onwards.
>
> So, I asked, how did the SSS system evolve? Martini replied:
>
>> The basic problem with a motorcycle equipped with telescopic forks is not the front suspension design, but the way its operation affects the frame design. In the CX the front suspension is still a fork, which is complete with pivoting steering head, yokes and hydraulic damping inside the suspension strut. This is, I consider, the next stage of what is accepted as conventional front suspension, after the inverted forks that are currently accepted as the leading edge of technology!

Created by the former Bimota engineer Frederico Martini, the CX125 had single-sided forks front and rear, allowing quick wheel changes to take place.

The CX Project

Martini went on to explain that the CX design had several advantages over even the inverted forks, which were fitted to Gilera's own SPO2/Chrono race replica series. This included a 13 per cent reduction in weight (much of it unsprung) accompanied by a massive 40 per cent increase in rigidity, reduced gyroscopic influence, central location along the longitudinal axis of both front and rear disc, providing more balanced braking effect, superior space in the area around the steering head, and simpler removal/replacement of the front wheel. This latter feature was also an advantage at the other end of the CX, thanks to the use of a single-sided swinging arm.

Whilst on the subject of suspension, it is worth noting that the suspension travel of 100mm front and 135mm rear was relatively short. However, when riding the motorcycle it was surprisingly stable and almost comfortable – remember that we're talking about a 264lb (120kg) 125 sports bike.

Compared to the usual racer-oriented engine tune of other leading edge 125s, the CX's power delivery was both smooth and torquey. It being possible to run down to as low as 3,000rpm in top (sixth) gear. Engine development engineer Pier Luigi Bertolucci said that 'some top end power has been sacrificed for superior low and mid-range output, but the slippery shape of the CX's bodywork largely offsets this deficit'.

The braking was also unusual, with both it and the special wheels being made especially for Gilera by the Grimeca concern. At the front was a powerful 300mm single disc with a four-piston caliper.

Both the riding position and the protection afforded by the all-enveloping fairing were truly superb, as was the comprehensive automobile-style dash, which was an example of how bike builders ought to do things; the CX125 put many bigger and far more expensive bikes to shame in this area.

But, however good the CX125 was, the concept really needed a larger, four-stroke power unit, or, for that matter, a series of engines, to make full use of both the technology and the unique style. But unfortunately this was not to be, as Gilera, or at least the Arcore-built motorcycle range, was stopped in its tracks with Piaggio's decision at the end of 1993 not only to close the plant, but also to axe motorcycle production altogether. We can only imagine what might have been …

11 Technical Appraisal

DEVELOPMENT

This chapter sets out to show the reader the main technical components associated with the classic four-cylinder Gilera engine that won several World titles during the 1950s, ridden by the likes of Umberto Masetti, Libero Liberati and Geoff Duke.

The following is a brief recap of the development history. The Gilera four had its origins in 1923. In that year two Rome-based engineers, Carlo Gianini and Piero Remor, designed and built a 500 transverse four with a single overhead camshaft driven by a train of gears set between two pairs of cylinders.

The following year, joined by another Roman motorcycle enthusiast, Count Gianini Bonmartini, they produced their first complete machine – the GRB (Giovanni, Remor, Bonmartini) – and development of this model continued up to 1928, by which time it was producing 28bhp at 6,000rpm.

An air-cooled unit up until then, the power unit was modified in 1927 by having the exhaust area of the portion around each sparkplug cooled by water.

GRB was eventually taken under the wing of the CNA organization (*Compagnia Nazionale Aeronautica*) of Rome (owned by Count Bonmartini), and Ing. Gianini became solely involved with engine design for light aircraft. Later still, CNA itself was taken over by the Caproni aviation company.

By now the four was known as the *Rondine* (Swallow), but Caproni were not interested, so the motorcycle design and rights were sold to Gilera. As the Rondine, it now had twin overhead camshafts, still with central gear drive; its inclined cylinders were fully water-cooled; and it was supercharged. Power output was a claimed 80+bhp at 9,000rpm.

Under its new name and ownership, the Gilera four enjoyed considerable fame and success in the years leading up to the outbreak of the Second World War – both in the fields of Grand Prix racing and World Speed Records.

In 1939/40 the Gilera racing department (Italy didn't enter the war until 10 June 1940) used experience gained with the 500cc machine as a basis for the design of an air-cooled 250cc four which had the supercharger mounted in front of the crankcase instead of behind the cylinders. Although never actually raced, it was this smaller engine that set the pattern for the post-war 500cc four.

Immediately after the war, Gilera raced the old water-cooled four, but the removal of the supercharger to conform with the FIM's newly introduced ban on superchargers dropped the power output by almost half. As it no longer had a competitive edge, it was decided to build a new 500 four which would incorporate the design advances featured in the 1940 250.

Teething troubles took up most of the late 1940s – the 500's first victory coming in the 1948 Italian Grand Prix.

The FIM introduced the new World Championship series in 1949 and Gilera's first title came in 1950 with Umberto Masetti at the controls.

Designed by Remor, the newcomer featured a unit construction engine with a four-speed gearbox producing around 47bhp at 9,000rpm.

Remor left to join rivals MV at the end of 1949, and development was continued by Ing. Columbo and Ing. Passoni. From then, until their withdrawal from racing at the end of 1957, the four clocked up a host of GP victories and World Riders' and Manufacturers' titles. A 350cc version was introduced in 1956. Finally, in November 1957, the factory secured a fantastic collection of World Speed Records, the most noteworthy being Bob McIntyre's incredible 141mph (227kph) for the One-Hour on a 350.

The fours made something of a comeback in 1963 under the watchful eye of Geoff Duke's Scuderia Duke squad; later still the fours were raced by the likes of Benedicto Calderella, Remo Verturi and Frank Perris. A major change, right at the end, was a seven-speed gearbox, which replaced a five-speed assembly that had arrived during the early 1950s.

TECHNICAL DETAILS

The four-cylinder Gilera engine was a truly wonderful piece of engineering and set the style in multi-cylinder design which has since been followed by several other Italian and foreign factories – notably the Japanese.

The main element of its crankcase was a single-piece electron casting that formed the lower half of the crankshaft chamber, the housing for the primary drive, oil pump, clutch and gearbox. The crankcase was then completed by the upper half of the crankcase chamber with its flanges for the four cylinders (horizontal splitting of the crankcase was required to fit and remove the crankshaft) and four separate covers. There was one of these detachable covers at each side of the crankcase chamber; one large cover on the clutch dome, on the nearside of the engine, permitted withdrawal and replacement of the clutch and gearbox internals, and another on the offside provided access to the rear drive sprocket.

The cylinders were inclined forward at an angle of 30 degrees to provide better cooling of the head and also to keep the engine height as low as possible. Various types of cylinder had been tried: light alloy with chromed bore, light alloy with cast iron liner and light alloy with chromed cast iron liner. The best results were those obtained with a cast iron liner.

The cylinder finning was not particularly extensive and, in fact, it had to be cut back between the cylinders in order to keep down the overall width of the engine – which was no more than 19.7in (500mm). However, there were never any problems with overheating, partly because of the high speeds at which the machines travelled, but also because of a long-stroke engine which meant that there was a greater proportional height of cylinder that could be finned. The actual bore and stroke dimensions were 52 × 58mm for the 500cc power unit (giving an individual cylinder displacement of 124.876cc, totalling 499.504cc) and 46 × 52mm for the 350cc (87.416cc per cylinder, 349.664cc overall).

Originally the cylinders were arranged in banks of two, but later each was separate, although the cylinder heads were in pairs. These were of light alloy with steel valve seats. The valves were inclined at an angle of 100 degrees (90 and 80 degrees having also been tried) and the exhaust valve, which had a 15 per cent smaller diameter head than the inlet, was sodium cooled. An eight-valve featured triple helical springs, enclosed by cylindrical tappets which were in direct contact with the cams.

The combustion chambers were perfectly hemispherical. To enable the use of the largest possible diameter valves, however, the 10mm spark plugs were not set right in the centre of the combustion chamber but in a small separate chamber linked with the combustion space by a narrow slot between the valves. This is a system which had been employed on other Italian racing engines and also on racing cars.

Various types of crankshaft were tested, these included: those that had been forged in one-piece, necessitating the use of split big-ends and split cages for the big-end rollers; those that had been built-up in various pieces united by press; and finally those that had also been built-up in various pieces united by the Hirth system. Some Hirth types built by the specialist German firm cost Gilera 3,000,000 lire each which, in 1957, was a pretty substantial slice of cash. However, it was found that these didn't really offer any real advantage over other types made in Gilera's Arcore factory using the best Italian steels.

Finally, it was found that built-up crankshafts gave the best service. These had a service life of between 50–100 hours under racing conditions and were binned even if no apparent wear could be detected. Similarly, all other highly stressed components were systematically changed at 50–100 hours. This echoed practices from the aviation industry.

The crankshaft had eight full-circle flywheels and ran in four inner roller bearings (with split cages) and two ball-race bearings, one at each end of the shaft. To keep engine width to a minimum, the second flywheel from the nearside had a toothed periphery, engaging with the outer drum of the clutch to provide the primary drive. At the centre of the crankshaft there was a timing pinion, driving the inlet and exhaust camshaft by means of a train of four gears, all running on ball-bearings. The complete timing gears were enclosed in an oil-bath case in the centre of the cylinders, the engine being notable for a complete absence of external oil pipes.

The crankshaft also drove the oil pump which drew its lubricant (a castor-based SAE 40 was used during its GP career) from the 1 imp. gal (5ltr)-capacity sump.

Drive for the Lucas magneto was taken off the clutch drum. Various other types of magneto were also tested, while battery/coil ignition was also experimented with, but at 10,500rpm for the 500cc unit and at 11,000rpm for the 350cc engine, the Lucas equipment proved the most reliable. Ignition timing was fixed to give an advance of around 60 degrees.

The original four-speed gearbox was replaced by a five-speed assembly. And, later still, during the mid-1960's, a seven-speed set was fitted.

There now follows a series of photographs that will provide the reader with the most noteworthy engine components of the four-cylinder Gilera racing engine.

Technical Appraisal

This view of the 499.504cc (52 × 58mm) long stroke four-cylinder engine provides a good view of not only the power-plant, but also the tubular steel, full-cradle duplex frame.

In the side view details such as the carbs, magneto, gearbox outer casing and massive oil sump are shown.

(Above) Frontal view of the engine showing finning for the sump, crankcase, barrels and cylinder heads.
(Below) Clutch components including outer pressure plate, centre and primary drive gears.

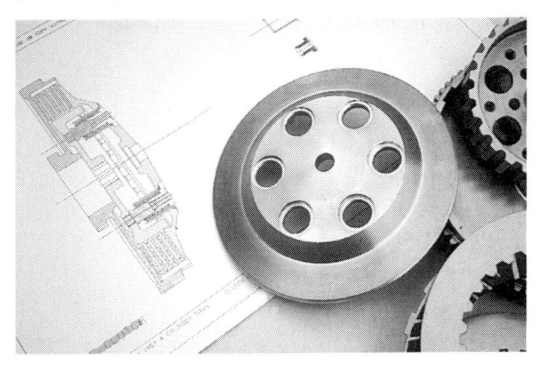

Friction and plain clutch plates, plus pressure plate, clutch centre and primary gear.

Technical Appraisal

(Left) *Technical drawing of the four-cylinder Gilera's clutch.*

(Below) *Crankcase vent components.*

(Below) *Gilera valve components, includes valves, guides, springs and tappet blocks.*

(Above) *Oil pump, crankshaft driven.*

(Below) *Exhaust header-pipe securing nuts.*

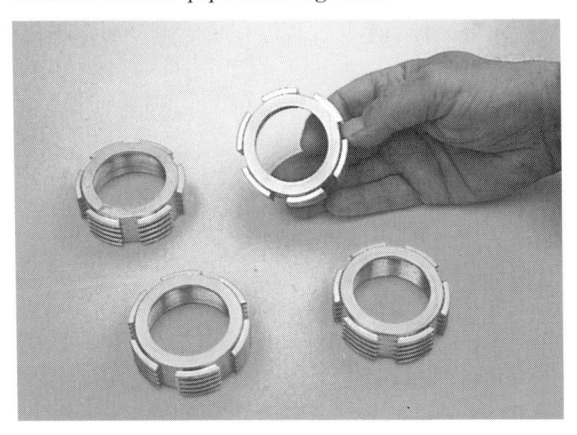

(Above) *27mm Dell'Orto SS carburettor from 500 Grand Prix four – mounted in pairs as shown.*

Technical Appraisal

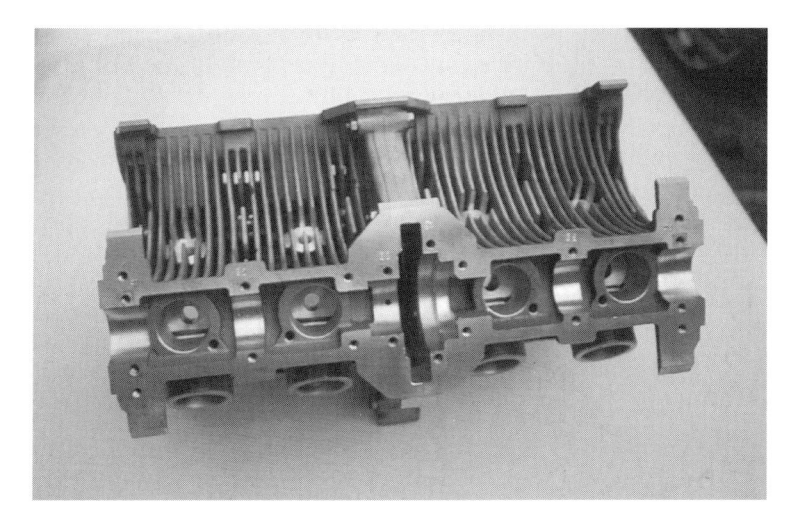

(Above left) *The central cam and timing gears for the cylinder head. These run in the opposite direction (cams go anti-clockwise) as compared to MV Agusta.*

(Above right) *Aluminium cylinder barrels with cast iron liner proved to be the best option.*

(Left) *Crankshaft head, upper view.*

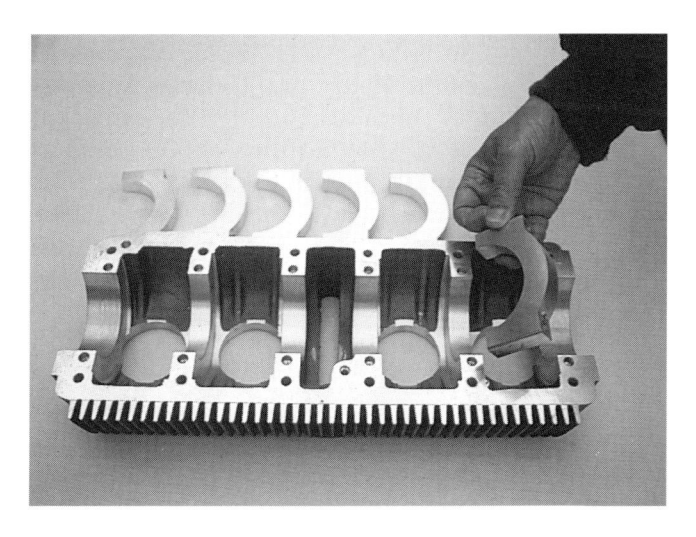

(Bottom left) *Crankshaft carrier and straps*

(Below) *Various types of crankshaft were race tested, but Gilera ultimately found the pressed-up type, as seen here, superior.*

12 Rebirth

For all those countless enthusiasts around the globe who thought the Gilera name had been lost at the end of 1993, when Piaggio announced that it would no longer be manufacturing the Arcore factory's two-wheelers, there was actually life after death.

To start with, unlike many former motor-cycling plants – for example Triumph's Meriden factory complex – Gilera's former Arcore home is still alive and well, albeit these days churning out car components for the Fiat group – and is still owned by Piaggio.

A PHOENIX FROM THE ASHES

Furthermore, Gilera itself has literally risen from the ashes, and is today producing a family of scooters and light-weight motorcycles.

The British arm of Piaggio, based in Orpington, Kent, has again begun bringing in Gilera-badged machines; starting in the spring of 1997 with two 50cc-class scooters, the liquid-cooled Runner and air-cooled SKP.

THE RUNNER

For the Runner, Piaggio claimed 'it combines for the first time, the practicality and ease of use of a scooter, with the rigidity, and handling of a fully-fledged motorcycle'. Whether this is actually true is open to question, with several machines from the past already attempting to bridge the gap between scooter and motorcycle, such as the Moto Guzzi Galletto, Velocette LE and Aermacchi Zeffiro to name but three. But what set the new Gilera apart was that earlier efforts not only dated from the 1950s, but, unlike the Runner, featured larger motorcycle-type wheels.

The Gilera Runner's specification included features such as a super-rigid frame construction, inverted (upside-

Gerhard Berger, 1995 Ferrari F1 team driver with his Gilera Typhoon paddock bike.

Gilera Runner 125/180: Scooters (1998)
[specifications for the Runner 125cc in brackets]

Engine type:	Single-cylinder, two-stroke, liquid-cooled
Bore:	65.6mm [Runner FX: 55]
Displacement:	175.8cc [Runner FX: 123.5]
Compression ratio:	9.8:1 [Runner FX: 9.9:1]
Max. power:	21/8,000CV/rpm [Runner FX: 15/7,500]
Torque:	19/7,000 Nm/rpm [Runner FX: 14/7,000]
Fuel:	Unleaded fuel ninety-five octane min.
Induction:	Lamellar valve in crankcase
Timing:	Piston-port
Carburettor:	Dell'Orto PHVB 20.5/Mikuni VM20
Choke tube diameter:	12mm
Ignition:	Electronic capacitive (CDI) and variable advance
Starter:	Electric and kick-starter
Lubrication:	Separate with automatic oil pump
Battery capacity:	12Ah
Electrical system:	12v
Gears:	Automatic CVT variator with torque converter
Final transmission:	Gears in wheel-hub (13/31 × 12/420) [Runner FX: gears in wheel-hub (12/32 × 12/42)]
Clutch:	Dry automatic centrifugal with muffler pads
Total short running ratio:	1/18.95 [Runner FX: 1/22.31]
Total long running ratio:	1/6.93 [Runner FX: 1/7.37]
Chassis:	Cradled in welded steel tubes with reinforcements in pressed steel
Front suspension:	Telescopic fork with upside-down shafts (30mm) and hydraulic shock absorber
Rear suspension:	Single-arm with dual effect hydraulic shock absorber and helicoidal spring
Front brake:	Stainless steel 220mm disc with hydraulic operation
Rear brake:	Stainless steel 240mm disc with hydraulic operation
Front wheel rim:	Die-cast aluminium alloy, 3.50 × 12
Rear wheel rim:	Die-cast aluminium alloy, 3.50 × 13
Front tyre:	Tubeless 120/70-12
Rear tyre:	Tubeless 130/60-13
Fuel tank capacity:	1.98imp. gal (9ltr)
Oil tank capacity:	0.4imp. gal (1.8ltr)
Max. speed:	74.5mph (120km/h) [Runner FX: 64.5mph (104km/h)]
Acceleration to 98ft (30m):	3.8 seconds [Runner FX: 4.2]
Acceleration to 197ft (60m):	5.4 seconds [Runner FX: 6.1]
Consumption (ECE cycle):	70mpg (4l/100km)
Max. length:	70in (1,780mm)
Width:	28in (720mm)
Saddle height:	32in (815mm)
Wheel base:	51in (1,303mm)
Moving weight:	253.5lb (115kg)

Rebirth

(Above) The Runner's up-to-the-minute instrumentation.

(Left) The 1997 Gilera Runner scooter with its liquid-cooled 50cc engine and automatic transmission. Later 125 and 180cc versions were added.

down) front forks, alloy wheels with low profile tyres, a powerful hydraulically operated disc front brake and a sleek fairing with flowing lines.

Other details of interest were 12in tyres, a centrally positioned fuel tank and filler, adjustable mirrors in fixed frames, and directable hot air vents; there was also the added advantage of a large dual saddle.

THE SKP

Even though it only had an air-cooled (as opposed to the Runner's liquid-cooled) power-plant, the SKP was classed as a sports scooter, and, like the Runner, featured automatic transmission. Braking was enhanced by the employ-ment of a large 190mm diameter front disc.

1999 Gilera Runner Trophy

For 1999 the Runner SP50 and 125 scooters provided the base for the Gilera Runner Trophy, a successful, single-make International Championship which involved more than 200 riders in over fifty races all over Europe. The best riders in each country raced for the European title in the grand finale of the series held over the Mugello circuit near Florence, on 9/10 October 1999.

The series was organized in two engine sizes (70 and 180cc) in collaboration with engine tuning specialists Malossi. The 70cc version producing 18bhp, whilst the larger Runner-based racer gave 26bhp.

The trophy events were run in the form of National Championships in Italy, Spain, Germany, Belgium and Austria.

AUTO ADVANTAGES

The advantage that both the Runner and SKP had as automatics was that it enabled their users to simply 'twist-and-go', nipping through urban traffic without the need for gear changing. Both models shared the same 40 × 39.3mm bore and stroke dimensions, giving an engine displacement of 49.3cc.

(Below) *Powerful and robust with its reinforced frame, 30mm upside-down hydraulic front forks and chunky tyres, the SKP50 scooter conquered city traffic and beyond.*

Gilera SKP 50 Scooter (1997)

Engine type:	Two-stroke, single-cylinder, air-cooled
Bore:	40mm
Stroke:	39.3mm
Displacement:	49.3cc
Compression ratio:	11.1:1
Max. power:	Depending on market
Torque:	Depending on market
Fuel:	Unleaded
Induction:	Reed valve in crankcase
Timing system:	Piston-port
Carburettor:	Weber/Dell'Orto 12mm
Ignition:	Electronic capacitive
Timing:	17 degrees
Start:	Electric kick-start
Lubrication:	Automatic oil pump
Battery:	4Ah
Electrical system:	12V
Gear:	Automatic CVT
Final transmission:	Gears
Clutch:	Centrifugal automatic
Short running ratio:	40.2
Long running ratio:	10.55
Chassis:	Welded steel tubes with pressed steel reinforcements
Front suspension:	Fork with oscillating shaft and dual-effect hydraulic shock absorber
Rear suspension:	Hydraulic single shock absorber, co-axial adjustable helicoidal spring
Front brake:	Stainless steel disc 190mm with hydraulic command
Rear brake:	Stainless steel disc 175mm with hydraulic command
Wheel rims:	Aluminium alloy 3.50 × 10
Front tyre:	120/90-10
Rear tyre:	130/90-10
Saddle height:	31in (790mm)
Fuel tank:	1.32imp. gal (6ltr)
Oil tank:	0.35imp. gal (1.6ltr)
Max. speed:	Depending upon market
Length:	69in (1,760mm)
Width:	28in (720mm)
Wheelbase:	48.5in (1,230mm)
Moving weight:	176lb (80kg)

THE SPANISH CONNECTION

Also being produced, and on sale in Europe from the 1997 model year, came the Eaglet, a 50cc liquid-cooled, custom-styled motorcycle with an electric starter, and the RK50, a 50cc trail bike with six-speed gearbox. The latter machine was built at Piaggio's Spanish arm, whereas the Eaglet and the scooter were all constructed at Pontedera, the main Piaggio plant in Southern Tuscany, near Pisa.

LARGER ENGINE SIZES

The success of the Runner led to a larger engine size being offered from the 1998-model year, both 123.5cc (55 × 52mm) and 175.8cc (65.6 × 52mm) now being available. This allowed Gilera to widen its potential customer base – and marked the brand name's return to larger engines, albeit something of a modest start.

THE CRUISER PROJECT

A prototype, concept bike if you will, was displayed on the Gilera stand at the German Cologne Show in September 1996. This was further developed and was put into production during late 1999, for sale in the 2000 model-year catalogue.

This machine, named the Cougar, is powered by an air-cooled, overhead cam, single-cylinder engine displacing 124.1cc (56.5 × 49.5mm).

Aimed at the lucrative European custom motorcycle market, the Cougar boasts a dual cradle frame in tubular steel, with a set of telescopic forks (35mm diameter stanchions) at the front and dual hydraulic dampers at the rear.

The 17in wheels in brushed aluminium feature stainless steel spokes, with a single 260mm disc at the front, and a 160mm drum brake at the rear.

Other details include: five-speed gearbox, five-plate wet clutch, capacity discharge (CDI) ignition and a dry weight of 271lb (123kg).

FINANCIAL MATTERS

In the summer of 1999, TPG (Texas Pacific Group), an American-based financial organization, was widely reported as being close to a deal that would see it take over the Piaggio empire. As TPG had already proved its ability in this field (in a buy-out of Ducati from previous owners Cagiva over the period 1996–1998), there were big hopes expressed by both Piaggio products in the two-wheel world, Vespa and Gilera.

However, in early November 1999, it was revealed that Morgan Grenfell, the merchant-banking arm of the German Deutsche Bank, looked set to take an 80 per cent share, leaving the balance with the original owners, the Agnelli family.

The Cougar custom bike with 125cc ohc single-cylinder engine, was first shown in prototype form during the 1996 Cologne Show. It entered production at the end of 1999 for the 2000 model year.

Rebirth

Gilera Cougar 125 Cruiser Motorcycle (2000)	
Engine:	Single-cylinder, four-stroke, air-cooled
Valves:	Two
Bore:	56.5mm
Stroke:	49.5mm
Displacement:	124.1cc
Fuel:	Unleaded petrol
Carburettor (diff):	22mm
Choke:	Manual
Ignition:	Capacitive (CDI)
Starting:	Electric
Clutch:	Five plates in oil bath
Front brake:	Single 260mm stainless steel disk with twin piston floating caliper
Rear brake:	160mm drum
Frame:	Double cradle of stainless-steel pipes
Front wheel:	17in aluminium with stainless-steel spokes
Rear wheel:	17in aluminium with stainless-steel spokes
Front tyre:	100/80
Rear tyre:	130/70
Front suspension:	Telescopic hydraulic 35mm, stroke 130mm
Rear suspension:	Large oscillating fork with twin hydraulic dampers, stroke 110mm
Max. length:	87in (2,219mm)
Max. width:	28in (720mm)
Max. height:	28.75in (730mm)
Dry weight:	271lb (123kg)
Fuel tank:	2.64imp. gal (12ltr)

Gilera's prototype Compact Bike made its début in 1999 – could this be a trend setter?

Piaggio in the late 1990s

A new range of superbikes could be on the cards now that outside investment has reached within the Piaggio empire.

Piaggio, owners of the Vespa, Puch and Gilera brand names, already had a range of four-cylinder Gilera sports bike engines on the test bench in 1998, but lack of cash had prevented them being taken any further. However, after interest was shown, first by TPG and then by Morgan Grenfell in late 1999, all this appears likely to change.

Outside investment also means a massive lift for Piaggio's workforce, who had struggled to cope with the death of chief executive Alberto Agnelli in December 1998, and then his mother Antonella Bechi Piaggio, early in 1999. Their deaths left the firm floundering and the remaining shareholders had been considering selling from that time onwards.

Piaggio is one of Italy's oldest industrial companies. It was founded in 1884 and has remained in the ownership of one family since that time. Scooter production began in 1946, and more than sixteen million Vespas have been sold since – making it Europe's largest two-wheel-powered manufacturer.

Bob Wright – Gilera Spares Specialist

Bob Wright Motorcycles began, as so many before, as an extension of a hobby, on 1 May 1975, in Western Super Mare on the south-west coast of England. Right from the very start it was a small operation which was to remain within the family's control – Bob running the shop, wife Jan doing the accounts. Their two sons helped in the early days after school and during holidays, until their own careers took precedence. Over the years since then the company has employed several mechanics, but a constant figure in the business has been Adrian Woods.

In the late 1970s Bob used to travel to nearby Bristol to pick up spares, notably from Douglas, the national Vespa and Gilera distributors. Then came a Gilera moped agency, followed shortly afterwards by the first advertisement in *Motor Cycle News* advertising Gilera spares – and so the first steps were taken in Bob Wright's subsequent rise to becoming the Gilera spares centre in Great Britain, a position it retains to this day.

Gilera enthusiast and spares specialist Bob Wright, with one of the ex-factory 500 four-cylinder racers at Imola in 1997.

Bob Wright – Gilera Spares Specialist *(continued)*

When Douglas closed in the early 1980s, Heron Suzuki took over, but their real interest was the Vespa scooter side, not Gilera. The result was the sale by Heron to Bob Wright of some four lorry containers of spares and bikes.

In a strange twist of fate Heron subsequently decided in the mid-1980s to import Gilera themselves, marketing the machines through London-based Suzuki racer/dealer Derek Loan. But this venture was not successful, so, once again, the unsold motorcycles and spares were purchased by Bob Wright. Later still, a large amount of spares from Kuala Lumpur, Malaysia were acquired, enabling the Weston Super Mare concern to cater for Gileras from the 1950s and 1960s.

To retain the stockholding which had been built up over the years, the next step was visits to Italy and these are now a regular feature of the business.

Bob Wright has also added to its range of parts availability by including both two-stroke and four-stroke spares from Gilera machines built during the late 1980s and early 1990s – including the Saturno and Nordwest.

Bob now says he is 'looking forward to catering for the new machines which are promised in the Millennium and continuing and extending the mail order business around the world'.

The Gilera Network

When news that the Arcore factory had been closed was released, the future seemed unclear. Owners in Britain were unsure if there would be any support or any dealers they could turn to. Owners needed some way of keeping in touch and passing on useful information. In 1994 Pete Fisher, David Champion and John Rushworth met at Donington Park to discuss setting up an owners' club. By the end of the day an objective and interim constitution had been agreed and John had also come up with the name – The *Gilera Network*.

Piaggio Ltd agreed to help launch the club by sending a mailing to all known Gilera owners. The Network quickly gathered about sixty members and in its first year was invited to join the Morini Riders' club camping weekend and track day at Cadwell Park.

In 1995 the Network became affiliated to the BMF and in 1996 Pete Fisher met Geoff Duke OBE. Geoff still wears a Gilera badge on his lapel and was keen to be associated with the Network. He kindly agreed to become a patron of the club.

Initially the Network attracted owners of new machines from Britain but, by 1996, the Network had members from all round the world and it continues to gather more classic Gilera owners.

The most important part of the Network is its quarterly magazine called *The GeN*. *The GeN* contains largely technical advice and tips to help owners run and maintain their bikes, but it also reports on the road-racing and hill-climbing adventures of several members. There are always plenty of letters and articles from members covering everything from tyre choice to touring. *The GeN* is doing what the club set out to do – keeping owners in touch and helping to keep Gileras where they belong – on the road.

The Cadwell Park weekend has become a regular feature and other events are always well supported. For example, in 1995 four members went to the TT where they announced that there would be a meeting at Glen Helen on Mad Sunday – nine Gileras turned up.

Bob Wright Motorcycles is Britain's only dedicated Gilera dealer and a spares lifeline for many owners around the world. Bob always has an advert in the magazine and is a keen Network member.

FUTURE PLANS

One of the major reasons for outside investment in the Piaggio group was to enable it to both increase production of its existing models, and also develop new models.

This was particularly true of the Gilera arm. By the late 1990s, brand name marketing was the buzzword. If proof was needed one only had to look at what this could mean in the bike business – Ducati, Harley-Davidson, Triumph and of course MV Agusta. Also, there were several new projects under development from other 'brand names' such as Benelli and Laverda.

With Gilera's history and the prestige of its four-cylinder racers, why not do an MV Agusta? The latter had seen an all-new, four-cylinder Superbike emerge in 1998 to a massive wave of publicity. However, reasoned the Piaggio board, to do this it needed a partner, hence the rush of interest mid-way through 1999.

So, as we enter the 21st century, it looks as if the likes of Ducati, BMW, Harley-Davidson and Triumph will be joined by another major name from the past in building a brand new Superbike for the new Millennium. Gilera is back, something that had seemed like an impossible dream only a few short years previously.

1997 Gilera Scooter Championship

Pioneered in Italy, scooter racing came to Great Britain for the 1997 Carnell Gilera Scooter Championship, organized by the MCRCB (Motor Cycle Racing Control Board) who had taken over the running of the British Championship road racing events from the beginning of the previous year from the ACU (Auto Cycle Union).

Norrie Kerr of Malossi UK was responsible for the preparation and supply of machines based around the Gilera 50 Runner scooter, but with a 70cc kit, which bumped the power up to around 16bhp.

Backed by Motorcycle Superstore chain, Carnells, the eight-round Gilera Scooter Championship was staged at Brands Hatch and Mallory Park (two rounds each), plus Snetterton, Castle Combe, Cadwell Park and Donington Park.

The winner was Leon Haslam, the thirteen-year-old son of former Honda works rider, Ron (Rocket) Haslam. Commentator Fred Clarke nicknamed Leon the 'Pocket Rocket'.

Index

Abarth 36
Aermacchi 79, 113
  Zeffiro 180
Agostini, Giacomo 79
Agusta, Count Domineco 58
Agnelli, Alberto 187
AJS 18, 45, 47, 52, 54, 65
  Porcupine 46
Aldrighetti, Giordano 81
Allan, Titch 119
Amm, Ray 90
Aprilia 144, 151
  Extrema 150
Aranda, Fernando 82
Argentine Grand Prix 77
Armstrong, Reg 48, 51–55, 57, 58, 60, 81, 82
Artesiani, Arcisco 29, 45–47, 81
Assen circuit 6, 45, 77
Avus circuit 60

Ballardini, Dario 145
Bandirola, Carlo 29, 43–46, 81, 124
Bartol, Harold 167
Bates, Vic 6
Belgian Grand Prix 45, 50, 58, 63, 81
Benelli 41, 48, 56, 63, 79, 189
Berger, Gerhard 180
Berne circuit 46, 67
Bertolucci, Pier Luigi 173
Bianchi 40, 73
Bimota 148, 149, 171
  Tesi 149, 171, 172
Blandford circuit 56
BMW 18, 23, 48, 52, 56, 57, 59, 80, 91, 169, 189
  GS enduro 161
Bologna Show 171
Bonmartini, Count Giovanni 78, 83, 112, 174
Bordeaux Grand Prix 34
Braglia, Claudio 145
Brambilla, Ernesto 40
Brands Hatch circuit 69, 73, 80, 182
Brough Superior 86
Brown, Bob 65, 67, 81
Brown, George 91

BSA
  Bantam 159
  B32 Trials 56
  Gold Star 159
Bucher & Zeda 10

Cadwell Park circuit 159, 160, 188
Cagiva 151
  Elefant 161
  Freccia 144
  Mito 144
Caldarella, Benedicto 6, 76–79, 81, 175
Caldarella, Salvador 77
Camathias, Florian 78, 81, 90, 91
Caproni 9, 83, 174
Carissoni, Pietro 68, 69, 124, 125
Carnell Motor Group 182
Casiraghi, Ing. Giovanni 113
Casoli, Paolo 167, 168
Castle Combe circuit 182
Cathcart, Alan 150, 153, 172
Cesena circuit 44
Champion, David 6, 188
Cheetham, Ann 159
Cheetham, Peter 159, 160
Ciancamerla, Romolo 167
Clarke, Fred 189
Clemencigh, Oscar 29
CNA 8, 9, 83, 174
C&N Enterprises 106, 109
Colnago, Giuseppe 32, 52, 54, 55, 58, 60, 82
Cologne Show 122, 184
Columbo, Alessandro 44, 45, 52, 175
Columbo, Luigi 79
Cremona circuit 10
Cruciani, Oliviero 166
Curt, Corrado 145
Cusi, Riccardo 145

Dale, Dickie 52, 54, 59, 81
Daihatsu 170
Daimler Benz 113
d'Ascanio, Ing. Corradino 112
Daytona circuit 76, 77, 90

Denegri, Gustavo 170
Deutsche Bank 185
Dixon, Freddie 115
DKW 41
Donington Park circuit 159, 182, 188
Douglas 114, 115, 119, 187
  Dragonfly 115
  Mark III 115
  Mark IV 115
  Mark V 115
Douglas, William 115
Ducati 68, 119, 149, 189
  250 racer 159
  800 vertical twin prototype 110
Duke, Geoff 6, 34, 47–49, 51, 52, 54–60, 62–65, 67, 69, 73, 74, 81, 82, 160, 174, 175, 188
Dundrod circuit 67
Dutch TT 16, 65, 75, 81, 89

ELF 149, 171

Faenza circuit 29, 44
Fairey 115
Fantic 144
FB Mondial 60, 63, 67
Fell, Todd 39
Fenocchi, Domenico 33, 124
Fernihough, Eric 86
Ferriari 167, 180
Ferri, Romolo 60, 61, 68, 69, 81, 82, 87–90
Fiat 36, 165, 167, 180
  Topolino 92
Fisher, Peter 164, 188
FN 24
Foell, R 78
Forconi, Tito 82
Francisci, Bruno 68, 82
French Grand Prix 54, 81
Frigerio, Ercole 49, 78, 82
Frith, Freddie 56
Fumagelli, Carlo 82
Fumagelli, Giovanni 62, 65, 74

Galliani, Claudio 82
Garelli 144

German Grand Prix 18, 57, 60, 69, 81
Giani, Martino 82
Gianini, Ing. Carlo 7, 8, 83, 86, 112, 174
Gilera – Dirt Bikes
  Saturno Cross 36, 37
  124 ohv Regolarita 102, 126
  98 ohv Regolarita 126
  125 ts Regolarita Casa 127
  175 ts Regolarita Casa 128, 130
  50 6V Competizione 118, 128, 129
  75 6V Competizione 118, 128, 129
  Elmeca Cross 130–132
  Elmeca Regolarita 128, 130, 131
  C1 Competizione Cross 131, 134
  C2 Competizione Cross 133
  E1 Competizione Enduro 131, 133
  125 Bicilindrica 132, 134
  HX 125 LC Enduro 135
  HX 125 LC Motocross 135
  HX 250 LC Motocross 135, 136, 139, 142
  NX 250 LC Motocross 136, 137
  RX 250 Rally 137
  Series K 134, 135
Gilera – Street Bikes
  317cc 10, 11, 102
  350 Super Sport 11, 20
  500 Super Sport 12, 20
  350 Gran Sport 13
  500 Sei Giorni 14, 20
  500 Tre Valvole 16, 20, 24
  175 Sirio 15
  250 LE 20
  250 L 20
  Marte 20, 22, 23
  VT500 Bitubo 20, 24, 26
  VT GS 20
  VTE 20
  VT GSE 20

Index

500 LE 20, 22
500 L 20, 22
Motocarro 20
250 Urano 20
500 Gigante 20
600 Gigante 20
500 LTE 22, 23
250 Nettuno 24, 37–39, 92
500 Saturno 10, 24–40, 45, 92, 110
500 Saturno Super Sport 36
Saturno Nuovo 155–160, 187
T Gran Sport 24
125 side valve prototype 92
125 ohv Turismo 93–97, 100
125 ohv Sport 93–97, 100
150 Turismo 95–97, 100
150 Sport 95–97, 100
250 Export 96–98
300 Extra 96–98, 106
B300 95–99, 101
175 Sport 100–102
175 Super Sport 100–102
175 Regolarita 100
175 Cross 100
Rossa Extra 101
98 Giubileo 102–104
125 Giubileo 102–104
175 Giubileo 103
202 Giubileo 104, 105, 122
G50 scooter 104–6, 108
G80 scooter 104–6, 108
Gilly moped 106, 107
Cadet 50 106
125 Sei Giorni 108
B50 5V 110
350 dohc twin prototype 110, 111, 115
500 dohc twin prototype 110, 111, 115
750 dohc three prototype 110, 111, 116
125 Arcore 115, 119, 120, 122
150 Arcore 115, 119, 120, 122
50 Trial 116, 117
50 Super 116, 122
50 Touring 116–118, 122
50 Enduro 116
50 RS 116
CBI 119, 143
CBA 119, 143
TS 50 119, 122
GR2 119
Trial 50 119

GR1 119, 121
TG1 119, 121
TG2 121
T4S 200 119, 122
TG3 121
T4 Custom 121
Eco moped 121, 145
Vale moped 121
GSA scooter 121, 143
R1 moped 143
125 RX 121, 138, 140–143
125 RWLC 121, 138–140
125 Arizona 121, 138, 139, 143
125 Arizona Hawk 139, 142, 143
125 RTX 139, 143
125 KZ 139–141, 143, 144
125 KK 139–141, 143, 144
125 RV 139, 142, 143
200 RV 139, 142
200 RTX 139, 142
250 NGR 139, 140, 142, 143
250 Arizona Rally 139, 142
125 KZ Endurance 140, 142
Fast Bike 143
MX1 143–146
Bullit 145, 170
RC Top Rally 145, 147
MXR 125 145, 146
XR1 145
XR2 145, 146
SPO2 145, 146, 149, 151, 171
MR2 146
Sioux 146
Apache 146
Crono 146, 147, 149, 151, 171
Freestyle 146, 147
CX 125 146–151, 171–173
SO3 146
Dakota 350 149, 152–157
Dakota 500 153–157
Nordwest 150, 160, 162, 163, 169, 187
XRT 600 155, 156, 158, 161
RC 600 161–163, 165, 168
RC 600 R 161–163, 168
GFR 150, 151
Runner scooter 180, 182, 183
Runner 125 181, 183

Runner 180 181, 183
SKP scooter 180, 182–184
Typhoon scooter 180
RK50 184
Cougar Custom 125
Compact Bike 186
Gilera – Racing Bikes
 Supercharged 500 four 15–19, 174
 Supercharged 250 four 41, 42, 174
 Air-cooled 500 four 42–82, 87, 91, 174–179
 Air-cooled 350 four 59–74, 82, 87, 91, 174–179
 125 dohc GP twin 41, 60–62, 68, 69, 82, 87–90
 175 dohc twin 68–70, 72, 82, 87–90
 250 GFR v-twin GP 166–169
 Saturno Bialbero 34, 35
 Saturno Corsa (Piuma) 31–34, 40, 67, 68, 82
 Saturno Competizione 28–31, 82
 Saturno Sanremo 29–32, 66, 82
 VT Corsa 24
 Nuovo Saturno Piuma 6, 156, 157
Gilera, Ferruccio 41, 60, 62, 63
Gilera, Giuseppe 10, 12, 41–44, 49, 62, 63, 67, 69, 78, 80, 83, 84, 108, 110, 170
Gilera, Luigi 12, 13, 15, 29, 49, 63, 85, 123
Gilligan, Gerald 6
Giro d'Italia 67–69
Graham, Les 45, 46, 52
Gramigni, Alessandro 168, 169
Grana, Rosseini 12, 123
Grassetti, Silvio 82
GRB 7, 8, 112, 174
Gritti, Alessandro 127
Guthrie, Jimmy 84

Hagiwara, N 157
Hailwood, Mike 75–78
Hansel, Peter 150
Harley-Davidson 189
Hartle, John 6, 69, 73–75, 81, 82
Haslam, Leon 189
Haslam, Ron 189
Hedemora circuit 53
Heron Corporation 115, 142, 158, 160, 169, 187

Hitler, Adolf 22
Hockenheim circuit 64, 67, 69
Hoffman 114, 115
Honda 63, 69, 74, 108, 145, 167, 169
 CB 750 6, 110, 111
 CBR 600 158
 CR 77 73
 CB 400 F 159
 Fireblade 169
 NSR racer 168
 350 K4 159
Honda Racing Corporation (HRC) 166
Hutchinson 100 74

Imola circuit 64, 65, 76, 78
Innocenti 114
ISDT 12, 13, 56, 103, 123–130
Isle of Man TT 45, 47, 52, 53, 55, 63–65, 73, 74, 79–81, 188
Italian Grand Prix 19, 35, 52, 56, 58, 60, 65, 78, 80, 81, 174
Itoh, C 155

Kawasaki
 GPZ series 142
Kay, Dave 6
Kay, Mark 6
Kelly, Sally 6, 157
Kerr, Norrie 182
Keller, Jacob 82
Klager, Ftiz 82

Lama, Francesco 82
Lambretta 114
Lavado, Carlos 166
Laverda 68, 144
Le Cap Rally 165
Lee, Nigel 163
Lerghy Frissel Hill Climb 164
Liberati, Libero 40, 45, 51, 54, 55, 59, 60, 62, 64–67, 71, 72, 81, 82, 174
Loan, Derek 187
Locarno circuit 56
Lomas, Bill 57
Lombardi, Giovanni 82
Lorenzetti, Enrico 50
Lotti, Carlo 145

Macchi, Felice 82
Maffers, Miro 12, 123
Magni, Arturo 45
Magri, Dino 78
Mallory Park circuit 6, 75, 157, 159, 182
Mancini, Franco 75

191

Index

Manx Grand Prix 34, 56
Martini, Frederico 147–151, 166, 167, 172
Martin, Leon 54, 82
Maserati 86
Masetti, Umberto 6, 47–52, 57, 81, 82, 174
Masserini, Massimo 24, 29, 44, 45, 82, 124
Masut, Ing. Lucio 152
Matchless 56, 65
  G50 77
McCandless, Crommie 81, 82
McIntyre, Bob 45, 62–65, 70, 71, 81–83, 175
Meani, Umberto 123
Meier, Georg 18
Mellone, Mario 24
Merlo, Ernesto 49, 82
Milani, Albino 49, 59, 78, 80, 82, 87–91
Milani, Alfredo 32, 47, 48, 51, 55, 58–60, 81, 82, 87–91
Milani, Gilberto 75
Milano–Taranto 14, 24, 33, 68
Milan Show 18, 92, 95, 96, 119, 121, 122, 135–39, 143, 144, 147, 149, 152, 154–6, 165, 167, 170
Mille Miglia 68, 87
Minter, Derek 6, 69, 73–75, 78–80, 82
Misano circuit 142
Modena circuit 24, 40
Monneret, Georges 82
Monneret, Pierre 52, 54, 60, 81, 82
Montlhéry circuit 86, 90
Montjuich Park, Barcelona circuit 32, 49
Monza circuit 11, 29, 35, 46–50, 52, 56, 58–63, 67, 69, 71, 73, 74, 78, 87, 89–91, 142, 150
Morgan Grenfell 185
Morini 68
Moto Guzzi 8, 22, 29, 48, 57, 63, 67, 86, 90
  Condor 24, 66
  Falcone 110
  Galletto 180
  700 V7 110
Motorcycle Racing Control Board (MCRCB) 189
Moto–Reve 10
Motor Imports 100, 108
Mugello circuit 142, 183

Mussolini, Benito 22, 41
MV Agusta 6, 33, 42, 43, 45, 52, 56, 58, 63, 69, 75–79, 86, 90, 95, 175, 179, 189

Nani, Dario 131
Napoli circuit 55
Nardo circuit 145
North West 200 56
Norton 8, 21, 22, 29, 33, 47, 51, 52, 54, 56, 59, 63, 69, 73, 80, 84, 90
  Manx 6, 21, 56, 67
  Seeley–Commando 159
  16H 21
  500T 56
NSU 65

Oldrati, Fausto 129
OPRA 8, 9, 41
Orioli, Eddi 150
Ospedaletti circuit 29
Oulton Park circuit 67, 74, 159
Ownsworth, John 159

Pagani, Nello 44–48, 51, 81, 82
Palermo circuit 24
Paris–Dakar Rally 137, 153, 156, 161, 168
Pasolini, Renzo 79
Passoni, Ing. Franco 34, 41, 42, 45, 47, 49, 51, 52, 54, 58, 59, 67–69, 71, 175
Patrignani, Roberto 72
Pattison, Shirley 6
Perfina, Maurizio 133
Perris, Frank 79, 80, 82, 175
Piaggio 63, 80, 111–122, 124, 151, 158, 165, 167–70, 172, 180, 184–89
  P32 aircraft 112
  P7 aircraft 112
  P119 aircraft 112
  P108 aircraft 112
  P136 aircraft 112
  P148 aircraft 112
  P149 aircraft 112
  Pegna aircraft 112
Piaggio, Antonello Bechi 187
Piaggio, Doct. Enrico 112
Pininfarina 144
Pio, Giovanni 34
Pochettino, H. Adolfo 82
Pride & Clarke 100, 108

Puch 165, 187

Read, Phil 6, 69, 74, 75, 82
Redman, Jim 74
Reims circuit 54
Remor, Ing. Piero 7, 12, 41–45, 47, 63, 84–86, 112, 174, 175
Ricotti, Ernesto 78
Rinaldi, Michele 131, 133, 134
Rivola, Luigi 145
Rocco, Silvio 145
Rolls-Royce
  Merlin aero engine 113
Rondine 8, 9, 12, 41, 48, 83, 112, 174
Ronson, Gerald 169
Rossetti, Amilcare 8, 9, 82
Rossi, Renzo 75
Rous, Charlie 73
Rover Group 169
Royal Enfield
  RE5 56
Ruggeri, Jader 29
Ruggia, Jean Phillipe 166, 167
Rusk, Walter 18
Rushworth, John 188

Sachsenring circuit 18
Saini, Gian Franco 124
Salmaggi, Ing. Giuseppe 24, 25, 110
Sambuy, Constantino 6
San Monica, Venezuela circuit 27
Sarolea 24
Scarborough circuit 159
Schneider Trophy 112, 113
Scuderia Duke race team 6, 56, 69, 73, 74, 76, 175
Serafin, Dorini 16, 18, 82
Silverstone 6, 69, 163
Snetterton circuit 182
Solitude circuit 57, 60
Soproni, Ernesto 29, 33
Spa Francorchamps circuit 67
Spanish Grand Prix 33, 35, 49, 78
Steyr–Daimler 165
Surtees, John 54, 58, 77, 87
Suzuki 56, 80, 187
  TS 125 139
Swedish Grand Prix 18, 53

Swiss Grand Prix 45, 46, 49, 51, 78, 81

Taruffi, Piero 8, 9, 12, 41, 46, 49, 52, 82–87
Texas Pacific Group (TPG) 185
Thorpe, John 95
Tonti, Ing. Lino 79
Tokyo Show 157
Tranchina, Giuseppe 6, 169
Tripoli Autodrome 8
Triumph 180, 189

Ulster Grand Prix 18, 19, 45, 47, 52, 54, 65, 81
United States Grand Prix 76

Vailati, Silvio 18, 82
Valdinocci, Orlando 32, 82
Vallelunga circuit 40, 79, 142
Varacca, Emilio 27
Varano circuit 142
Veer, Marc 54, 82
Velocette
  LE 180
Venturi, Remo 69, 76, 79, 80, 82, 175
Vergani, Fausto 126
Vespa 114, 115, 158, 165, 169, 187
Vezzalini, Enzo 82
Vianson, Enrico 115
Villa, Ercole 48
Villa, Ettore 24, 82
Vincent HRD 91
Viney, Hugh 56
Vintage Motor Cycle Club (VMCC) 119
Voice, Harry 34

Walker, Gary 159
Welsh, Ian 6
Westinghouse Group 115
Witteveen, Jan 131, 132, 135
Woods, Adrian 187
Wright, Bob 187, 188
Wright, Jan 187

Yamaha 167
  GTS 171
  Tenere 153

Zanchetta, Gino 123
Zeller, Walter 52, 57
Zündapp 23